The Power of Micronutrients – Milestones in Medicine and Health

PROF. DR. ELMAR WIENECKE

THE POWER OF MICRONUTRIENTS

MILESTONES IN MEDICINE AND HEALTH

Meyer & Meyer Sport

British Library of Cataloguing in Publication Data
A catalogue record for this book is available from the British Library

Original title: *Mikronährstoffe – Meilensteine der Gesundheitsmedizin*, © 2021 by Meyer & Meyer Verlag

The Power of Micronutrients
Maidenhead: Meyer & Meyer Sport (UK) Ltd., 2022
ISBN: 978-1-78255-223-9

Aachen, Auckland, Beirut, Cairo, Cape Town, Dubai, Hägendorf, Hong Kong, Indianapolis, Maidenhead, Manila, New Delhi, Singapore, Sydney, Tehran, Vienna

 Member of the World Sport Publishers' Association (WSPA)

Printed by Print Consult GmbH
Printed in Slovakia

E-Mail: info@m-m-sports.com
www.thesportspublisher.com

CONTENTS

PROLOGUE

by Prof. Dr. Elmar Wienecke

Health and individualization are among the megatrends that have fundamentally changed society. Health and individualization have become personalized medicine. The mindful handling of one's own resources and an increasing awareness of one's individual life energy requires new personalized health solutions. Trend researcher and futurologist Matthias Horx, whose *Zukunftsinstitut* ranks among the most influential think tanks in European trend research and futurology, provides content for the term "megatrends". According to the *Zukunftsinstitut*, the health megatrend focuses on how health can be improved and maintained at a younger age (https://www.zukunftsinstitut.de).

Recent findings show that every other German citizen, regardless of age, suffers from various disorders such as chronic fatigue syndrome, frequent infections, difficulty concentrating, lack of motivation, headaches, exhaustion, and much more.

The manager is burned out, the woman with the typical triple burden of "job, children, household" feels overwhelmed, and the retiree or pensioner suffers from a number of disorders. Far too often, therapists are unable to provide an adequate explanation for the causes.

> *"Human beings does not fall ill because the body lacks medicine, but because of biochemical disturbances in the body that are not detected and corrected in a timely manner." (B. Kuklinski)*

The research we have conducted in recent years show a subclinical undersupply of vital micronutrients in test subjects which, among other things, has resulted in many adverse effects on mental and physical performance capacity.

Smith and Myers (2018) of Harvard University in an extensive study (Impact of anthropogenic CO^2 emissions on global human nutrition) describe the loss of micronutrients in the vitamins, minerals, and trace elements in our diet due to the "growing greenhouse effect". Additional losses result from processing and storage of foods.

In recent years, we have analyzed and described the measurable losses and SALUTO (competence center for health and fitness in Halle/Westphalia), and with the aid of information scientists, worked meticulously for several years to integrate more than 60,000 case reports into a computer matrix. The database contains 10,542 target values or reference values of micronutrient concentrations in 297 categories with the corresponding cluster (gender, age, preexisting conditions, exercise habits) and is in line with results of long-term empirical research projects and studies. We were able to assign additional indicators for micronutrient formulations that are customized for each individual via a differentiated lab analysis. With the aid of the anamnesis questionnaire, results from special blood and urine tests, and the cluster classifications, the software uses specially developed algorithms to calculate the individual micronutrient formulation or rather the individual's requirement. This "individualizer" is a globally unique "evidence-based software" suitable for the preparation of personalized micronutrient formulations to maintain good health and to treat illnesses, without side effects.

THE POWER OF MICRONUTRIENTS

A quote from the periodical Meine Gesundheit (My Health) from the Foundation for Health and the Environment (SfGU) in Switzerland regarding the aforementioned database (Poeggler, 2017):

> *"A milestone in health care medicine – The globally unparalleled database opens up new possibilities in micronutrient medicine. The regulatory medical approach represents innovative successful health care medicine. It supplements and expands conventional medicine. As a pioneer in this area, Prof. Dr. Elmar Wienecke as the representative of his team, was awarded the innovation award for outstanding research and developmental work in the area of micronutrient therapy/regulatory medicine by the SfGU. A master program initiated by him at the FHM in Bielefeld/Germany will greatly contribute to improving the population's healthcare."*

The targeted supply of deficient micronutrients and the Glycoplan as per Dr. Mosetter have proven helpful in balancing these micronutrient deficiencies, i.e., biochemical disturbances.

Harness the power and strength of personalized micronutrient formulations for your fitness and quality of life, regardless of your age, at every age.

Sincerely,

Prof. Dr. Elmar Wienecke

You can find a personal message at:

PREFACE

By Andreas Hefel

No flimsy compromises when supplying micronutrients!

An adequate supply of micronutrients is essential to maintaining or restoring health and performance capacity. This fact has been sufficiently scientifically verified and is generally accepted. However, the key question of what maximum amounts of vitamins and minerals should be valid in nutritional supplements has been the source of controversial political debate for years.

Proponents argue that legal regulation would protect consumers from possible overdoses of certain substances by setting official upper limits. But in Europe an agreement on standardized maximum levels of micronutrients in nutritional supplements has so far failed due to varying risk assessments. One way out of this dilemma might be a change of perspective that would focus on the individual benefits to the consumer.

What is the desired effect of vitamins, minerals, trace elements, and secondary plant substances (antioxidants) on the metabolism? From a biochemical point of view, only the correct amount of the right substances ensures unimpaired metabolic regulation and thus health.

Nutritional supplements purchased, for instance, at the supermarket do not achieve that purpose. Such mono- or standard formulations are at the center of a political push regarding maximum amounts for vitamins and minerals. The German Federation for Food Law and Food Science (BLL) writes unapologetically about their benefits (https://www.lebensmittelverband.de/de/lebensmittel/nahrungsergaenzungsmittel/faq-nahrungsergaenzungesmittel):

> *"Even if nutritional supplements are offered in medicine-like form such as tablets, pills, or capsules, they are still food items and not medicines. They help maintain a sense of wellbeing. They do not cure, relieve, or prevent illnesses or health problems. That is also why they cannot and should not have to undergo a drug approval process."*

Thus, these products legally fail to address the actual needs of millions of people, namely to do something for their health the natural way. Therefore, if universal maximal amounts for vitamins and minerals must be specified, a properly functioning metabolism must not be the objective.

For this reason, customized micronutrient blends should not be compared to conventional nutritional supplements from the food sector. It is necessary to combine vital essential nutrients in such a way and at such high individual doses that the individual requirement is optimally met. The substances are ingested in the form of a precise dose that the person actually requires for good health. Nothing happens until deficiencies or shortages in the micronutrient supply have been accurately ascertained via laboratory findings – not based on suspicion or on a whim.

According to this insight there must not be any universal standards or maximum levels but rather the sum of individual ideal levels that make up the individual metabolic status. If the biochemical parameters determine that the metabolism is functioning smoothly the individual has an optimal supply. Only when the effectiveness of a custom blend cannot only be felt but also be measured, are we on the right track.

After thousands and thousands of laboratory analyses, it has become abundantly clear that micronutrient deficiencies are increasing. Here we can see a definite trend: On the one hand, an increased pollution burden on all levels amplifies degenerative processes.

THE POWER OF MICRONUTRIENTS

This continuously raises the micronutrient requirement. On the other hand, foods contain increasingly fewer micronutrients due to industrial processing and a higher CO^2 content, which causes plants to grow increasingly faster.

The results verifiably show that the gap between regenerative and degenerative forces is growing increasingly wider. That is also the reason why in industrial countries the number of healthy years people experience is shrinking. In Germany, for instance, that number is merely 57 years. At the same time, chronic illnesses, which are largely deficiency diseases, are on the rise. With that said, debates about maximal amounts of vitamins and minerals in nutritional supplements not only fall short, but completely miss the mark when it comes to health.

Treating the causes of deficiency diseases requires quality micronutrients as per the pharmacopeia, and can be mixed according to individual requirements. This approach can eliminate the risk of both an under- and oversupply.

As a pioneer in this field, Prof. Dr. Elmar Wienecke was given an award by the Foundation for Health and the Environment (SfGU) for his outstanding research and developmental work and its practical application. At the 12th International Conference at Lake Constance in 2017, he was awarded the innovation award, which included 10,000 € in prize money. He has also set another milestone as the initiator and chief scientist of the first master program in micronutrient therapy & regulatory medicine at the FHM University of Applied Sciences in Bielefeld/Germany. The regulatory medicine that has been taught there since October 2017 is generally accepted teaching also in and for conventional medicine and can serve as a certificate program for physicians' training, continued training, and advanced training.

The content is based on a solid and reputable scientific foundation. This makes it possible for the study program to significantly contribute to the public's health care and specifically prepare students for the challenges of actual practice. It opens up new perspectives in prevention, therapy, and rehabilitation. This book offers impressive documentation of how it can be done and the great successes that can be achieved.

Andreas Hefel

President, Foundation for Health and the Environment (SfGU)

Berlingen, Switzerland, June 2020

STATEMENT

By Dr. med. Kurt Mosetter

Crystal-clear results: Energy and performance are measurable and edible, and monitoring: Measuring, balancing, testing

The future of performance and regeneration optimization in sports will reach new horizons via metabolic training and the integration of the current body of knowledge about the microbiome and the metabolome, the energy metabolism, gut, liver, mitochondrial and brain axis. In doing so, smart nutritional medicine, food factors, and supplements will be just as critical as training methods.

Next to the principles of macronutrient regulation, micronutrients will play an essential role. Prof. Dr. Wienecke pioneered measuring and compilation of individual micronutrient analyses in which detailed amino acid and fatty acid profiles, mineral and vitamin spectra in whole blood, as well as differentiated testing of the individual steps of the citric acid cycle, can be made clearly visible. It is the reason Prof. Dr. Elmar Wienecke and his institute SALUTO already possess a reference database of more than 60,000 differentiated cases. Moreover, individual measuring parameters and their – from a modern point of view – often too broadly defined limits are no longer critical; instead, the more narrowly defined metabolic patterns are.

Personalized and customized vital substance measurements can optimize the performance spectrum and regeneration of pro athletes via very detailed measurements and precise analyses, and it can do the same for executives as well. Additionally, and particularly impressive: Even patients with complicated illnesses can benefit enormously from these differentiated micronutrient analyses; deficiency states and asymmetries below the surface of normal blood panels can be unmasked and paths to causal solutions can open up. Last but not least, circumscribed deficiency states can be caught at a time when classic small blood panels do not yet show changes. Early detection of undersupplies and disturbances of homeostasis makes real prevention measurable and edible.

When dysfunctional patterns are detected early enough, multiple potential illnesses can be intercepted and avoided while on the horizon, essentially "preempting the disease".

The basis for every smart supplementation is the evolution-derived "natural eating" nutrition management concept Glycoplan – only the best of everything! Strengthening the energy metabolism, particularly the turntables of the ATP-metabolism, the mitochondria with SCFA, MCT, coenzyme Q_{10}, NADH, creatine, ribose, and galactose, can stimulate the processes of regeneration and repair, protein and fasciae synthesis, real breakdown, customized recycling and new synthesis.

It is up to us to promote the regular operation of our molecular repair shops. Along with a sufficient amount of energy, our molecular repair shops require a well-appointed toolbox.

Amino acids like leucine, valine, isoleucine, Tyrosine, tryptophan, arginine, methionine, cysteine, lysine and proline, zinc, magnesium, selenium and chrome, Omega-3 fatty acids, all vitamins, especially Vitamins E, D, A, K_2 and B vitamins as well as the assurance of molecular signal carriers such as monosaccharides, will be critical. The latest scientific findings support the science that prevention and provisions are possible. Within the context of farsighted training management, injury prophylaxis, and regeneration, but also in the case of injuries, and rehabilitation, multiple accompanying avenues and different mechanisms should be pursued.

THE POWER OF MICRONUTRIENTS

In closing one more sentence: Thank you, dear Elmar! Along with our actually calculable successes in pro sports and pro soccer as well as regular steps towards recovery of severely ill patients, your successful efforts and accomplishments in research and teaching of the master program are outstanding. Its teachability and elegant practical applicability establish this methodology for a healthy future.

Dr. med. Kurt Mosetter [1]
Constance, Germany, August 2020

1 Dr. Kurt Mosetter is the director of the Center for Interdisciplinary Therapies, ZIT, in Constance, initiator of Myoreflex therapy, as well as the former physician for the US National Soccer Team, long-time advisor for the coaching staff at Hoffenheim and RB Leipzig, as well as an internationally sought-after advisor and physician of executives and elite athletes.

1 THE FUNDAMENTALS OF AILMENTS AND MICRONUTRIENTS

1 THE FUNDAMENTALS OF AILMENTS AND MICRONUTRIENTS

1.1 The vicious cycle: Circulus vitiosus

Every other German citizen, regardless of age, complains about various ailments. They are chronically fatigued, suffer from frequent infections, have difficulty focusing, suffer from lack of motivation, headaches, exhaustion, and much more. For many Europeans it is becoming increasingly difficult to escape this vicious cycle (fig. 1).

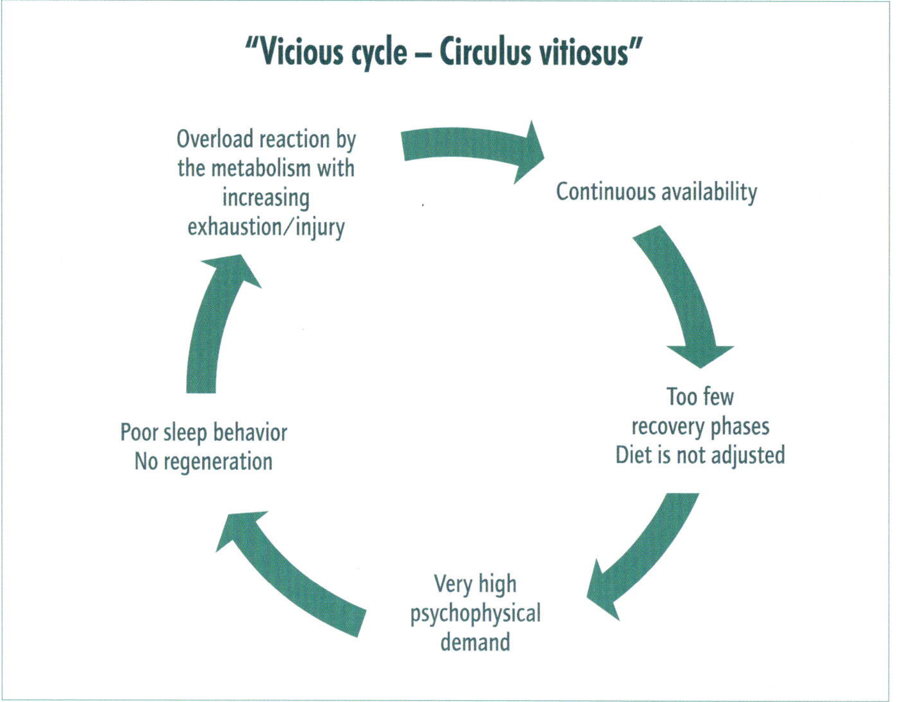

Fig. 1: Circulus vitiosus

Due to this demographic development, the pharmaceutical industry is experiencing a boom: The increasingly aging population and the seemingly inextricable link to polypharmacy, meaning the increasing consumption of different medications that often treat primarily the symptom but not the cause, are to blame. In fact, the science has verified that many health problems can be easily treated by detecting and correcting biochemical disruptions early. The targeted intake of insufficient micronutrients harmonizes or rather normalizes the necessary metabolic processes.

In recent years, more than 4,500 individuals who were taking prescription antidepressants have come to us at SALUTO (competence center for health and fitness in Halle/Westphalia). Our special laboratory analyses continuously show serious deficiencies in the area of specific amino acids (phenylalanine, tryptophan, Tyrosine), striking deficiencies in the area of cellular B-vitamins, and the beginning of thyroid dysregulation, which in some cases has caused significant anxiety in those affected. With the targeted intake of the insufficient micronutrients, the science shows that these disorders can be corrected after approx. 12 weeks (➡ chapter 6.3.5).

We could retrospectively show in 1,250 individuals with severe rheumatic disorders and an existing metabolic syndrome how an average intake of 8-10 medications could be reduced by up to six medications after 24 weeks through a change in diet and an added customized intake of lacking micronutrients. (➡ chapter 6.3.2)

Based on our results, the following postulate by the Federal Institute for Risk Assessment (https://www.bfr.bund.de/de/a-z_index/mineralstoffe-5074.html)

> *"In Germany an undersupply of vitamins and minerals is very rare in healthy people with a varied diet."*

is a farce, even if it is supported by the German Nutrition Society and other regulatory institutions (Bechthold et al., 2012a; Bechthold et al., 2012b).

THE POWER OF MICRONUTRIENTS

As far back as 2015, the results of the study *Global Burdens of Disease* published in the *Lancet* were debated in the publication *Die Welt*. The bottom line: Only one in 20 people is truly healthy (Global Burden of Disease study, 2013; Collaborators, 2015; Liebram & Nauber, 2015). Everyone else suffers from five or more simultaneous disorders and takes medications that result in the increased excretion of vitamins and minerals. Only 13.1% of the population have a fairly healthy diet. This implies an increasing undersupply of micronutrients.

Every third person older than age 50 takes at least 4-5 medications. The micronutrient deficiency triggered by pharmaceuticals is being intensely debated in the literature (Gröber, 2015; Gröber, 2018). This micronutrient deficiency resulting from pharmaceuticals leads to additional disorders and more medications are administered – a Circulus vitiosus, a vicious circle, is created.

Some readers may think that a little stress here and there isn't such a big deal. That's true, but it does decrease the health span.

What is the health span?

The *health span* offers information about the number of years a person is healthy. In Germany that time span lies at 57.4 years, which means that women and men, with an average life expectancy of 79.7 years, must live for years with significant health restrictions (fig. 2).

According to a statistic that was introduced at the Child and Adolescent Conference in Berlin in 2019, today every fourth child is already taking medications regularly to better cope with everyday stressors. This means that even childhood years cannot as a rule be included in the health span. The effects of micronutrient deficiencies can already be seen in children.

The Foundation for Micronutrients, Prevention, Health, and Quality of Life had the data from SALUTO scientifically analyzed as part of the master's program. The evidence-based retrospective study (➡ chapter 6.3.1) completed with 162 children who suffered from attention deficit/hyperactivity disorders shows the unbelievable effects that can be achieved with a targeted intake of insufficient micronutrients and a change in diet as outlined in Dr. Mosetter's Glycoplan (➡ chapter 4.7).

Life expectancy in Germany: 79.9 years
(Men: 77 years; women 82.4 years)

Health span: 57.4 years
(Men 57.1 years; women 57.7 years)

This means: 20 to 25 years
- Illness
- Suffering
- Pain

Fig. 2: Health span and life expectancy

This highlights that health span and an adequate micronutrient supply are inextricably linked.

1.2 Typical disorders

More fairytales and myths from nutritional medicine, such as the German population's aforementioned good micronutrient supply, will be discussed in the following chapters (➡ chapter 4.2). The insufficient energy balance of more than 14,000 test subjects on a cellular level correlates with massive micronutrient deficiencies that are linked to a variety of disorders. By contrast, routine medical exams of these same subjects did not reveal any deficiencies that appeared to be the cause of the disorders.

Micronutrient deficiencies can cause the following disorders:

- Poor sleep behavior;
- Increasing exhaustion;
- Lack of mental vigor;
- Poor stress tolerance: inner composure is quickly lost;
- Increased mood fluctuations;
- Inner unrest;
- Frequent night sweats;
- Increasing difficulty concentrating over the course of the day;
- Muscle tension, as well as
- Pain.

1.3 Micronutrients: What is it?

Nutrients are basically substances humans need for normal development and for maintaining good health. Nutrients can be divided into two groups:

- **Macronutrients:** These include carbohydrates, proteins, and lipids (fats).
- **Micronutrients:** These substances are necessary for intracellular utilization of macronutrients. They include vitamins, minerals (e.g. calcium, magnesium), trace elements (e.g. iron, zinc, selenium, manganese), secondary plant substances (carotenoids, flavonoids), essential fatty acids (e.g. fish oils) and amino acids.

Micronutrients must be ingested with food without supplying energy on their own. Foods that are rich in micronutrients are fruits and vegetables, but only in a freshly-harvested state. Storage and cooking destroy them. The result is that, contrary to the German Nutrition Society's postulate, the intake of micronutrients is most often inadequate.

Micronutrients have an antioxidant effect and serve to synthesize macromolecules or as cofactors for essential enzyme reaction.

1.4 Energy status principles

How are nutrients and the human body's energy status related? Energy can only be produced when the body uses micronutrients to convert macronutrients into energy via oxidation (➡ chapter 2.3) in order to:

- Maintain body temperature;
- Ensure physical and mental functions;
- Ensure growth, and
- Perform metabolic activities.

Energy demand differs from person to person and from day to day and is a result of

- Resting metabolic rate: The average amount of energy a person at complete rest and while lying down requires to maintain body temperature and basal metabolism (heart function, respiration, etc.), as well as
- Total energy expenditure, which results from all additional energy-consuming activities, such as growth, movement, digestion.

Energy demand and available nutrients determine energy status. When demand and supply are balanced, the energy status is equalized. This energy balance affects the individual's sense of wellbeing.

2 INNOVATIONS REGARDING MICRONUTRIENT REQUIREMENT

2 INNOVATIONS REGARDING MICRONUTRIENT REQUIREMENT

2.1 How high is the micronutrient requirement?

The current medical approach to assess micronutrient requirements is not very innovative:

Independent laboratories interpret the assessment criteria of possible deficiencies for the individual micronutrients differently. The references are not very valid since only a limited number of people were examined and most labs don't have sufficient information about the individual disorders of the examined persons.

Nevertheless, the labs provided the treating physicians with the respective valuations for the standard range of values. The range of values of what is healthy and what isn't are defined irrespective of the affected person's clinical status, and thus arbitrarily reported based on purely biochemical tests.

European DACH associations gather the results from the laboratory facilities and continue to promote the idea that a balanced diet can cover an individual's micronutrient requirements, without having sufficient comparative data from humans. Unfortunately, all of the DACH associations' data are based on routine serum tests and are not separated into differentiated specific clusters. When looking at Germany's seniors, whose vitamin intake as per DACH recommendations was identified as being too low, we can see serious deficiencies on the serum level (fig. 3).

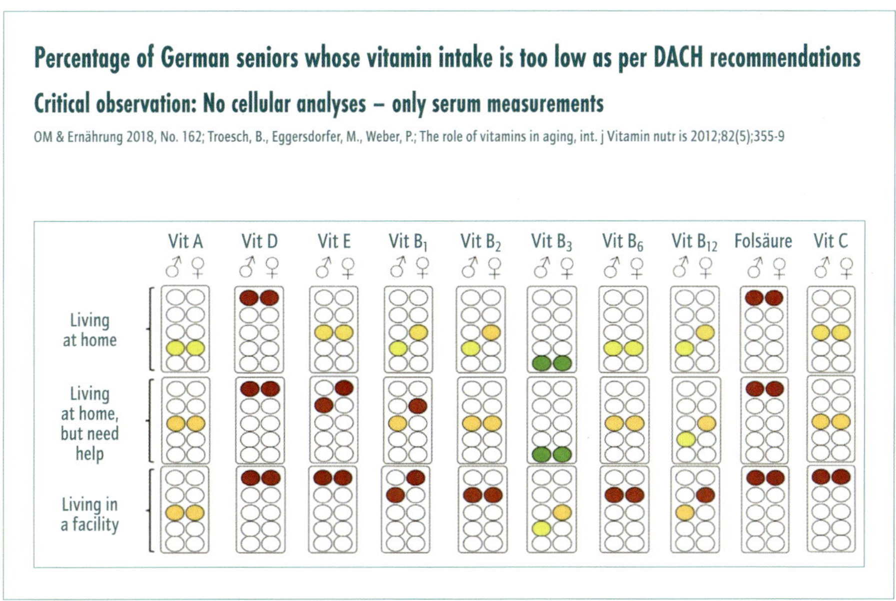

Fig. 3: Percentage of German seniors whose nutrient intake is too low as per DACH recommendations

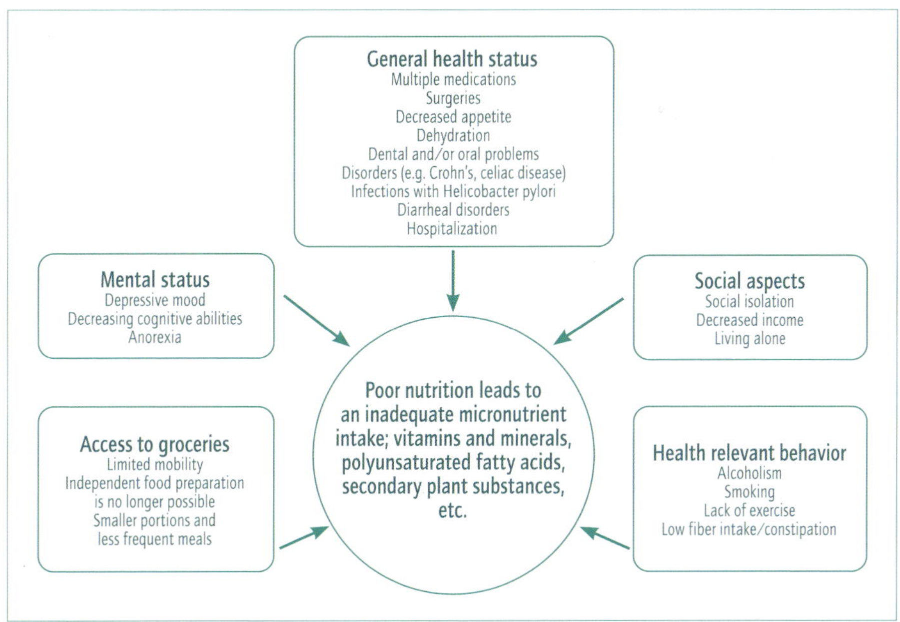

Fig. 4: Multiple factors that contribute to a poor micronutrient supply in older people.

THE POWER OF MICRONUTRIENTS

It is not surprising that the micronutrient supply in older people is deficient. Figure 4 shows the many factors responsible for increasing micronutrient deficiencies in our population as people get older.

Figure 4 also illustrates that, due to these many influencing factors, it seems plausible that the number of people with different disorders and preexisting conditions attributed to a lack of micronutrients is far greater than previously thought. The correct micronutrient diagnosis is the Alpha and Omega (➡ chapter 2.3) for a lasting effect on these disorders. Here a paradigm shift in lab diagnostics is urgently recommended (➡ chapter 2.3.2).

In recent years, the Foundation for Micronutrients, Prevention, Health, and Quality of Life, with support from SALUTO (competence center for health and fitness) and the FHM technical college, has been able to facilitate the extra-occupational masters and certificate program Micronutrient Therapy and Regulatory Medicine for physicians, physical therapists, pharmacists, nutritionists and sport scientists, and natural health professionals.

This course of study is based on many years of evidence-based studies that have been compiled in a comprehensive and globally unparalleled database with 10,542 target- and reference values of more than 60,000 tested individuals divided into 279 categories with a corresponding cluster (such as gender, age, preexisting conditions, athletic activities).

These evidence-based studies make it possible to present the effects of micronutrient therapy on different disorders and existing underlying medical conditions as well as in competitive and elite athletes (➡ chapter 6.3.8). The correct dosage of lacking micronutrients can verifiably optimize especially the vegetative nervous system and is scientifically proven to improve people's "sense of wellbeing" in the long-term.

Over the past 25 years, SALUTO's test results were able to show what the population's micronutrient status, and thereby its energy status, looks like, and what definite effects can be achieved in people via a targeted intake of lacking micronutrients ("Energie auf Rezept" (Energy on Prescription); ➡ chapter 6.2; fig. 5). Every individual, every single case, is documented and can be tracked scientifically in detail and is in line with the criteria of evidence-based medicine.

This allowed us to ascertain the status quo of the energy balance and micronutrients of 6,150 executives and 8,120 additional employees during the initial intake examination. After a 12-week change in diet and the customized intake of insufficient micronutrients, the many disorders were verifiably reduced (fig. 6).

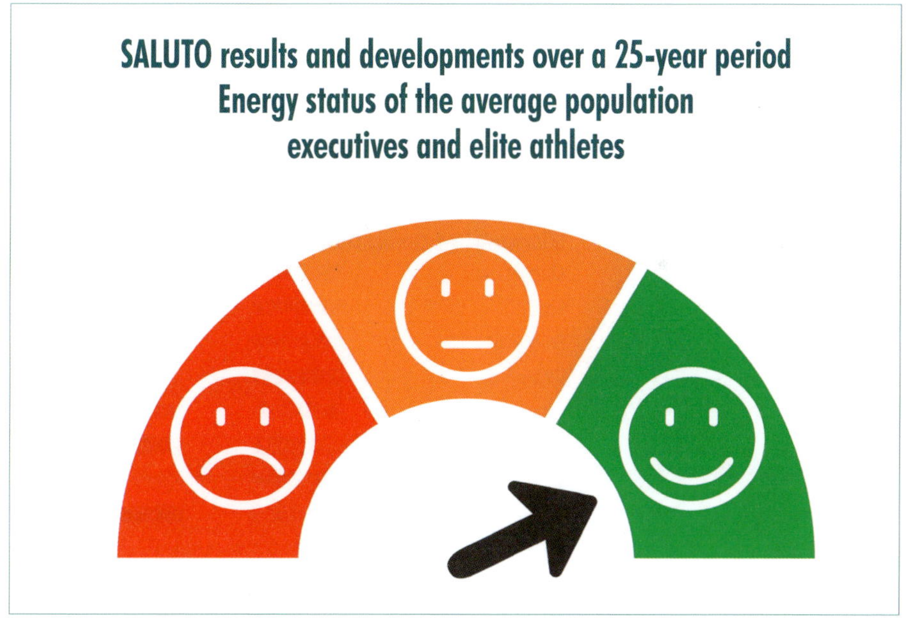

Fig. 5: Results and developments of SALUTO over a 24-year period

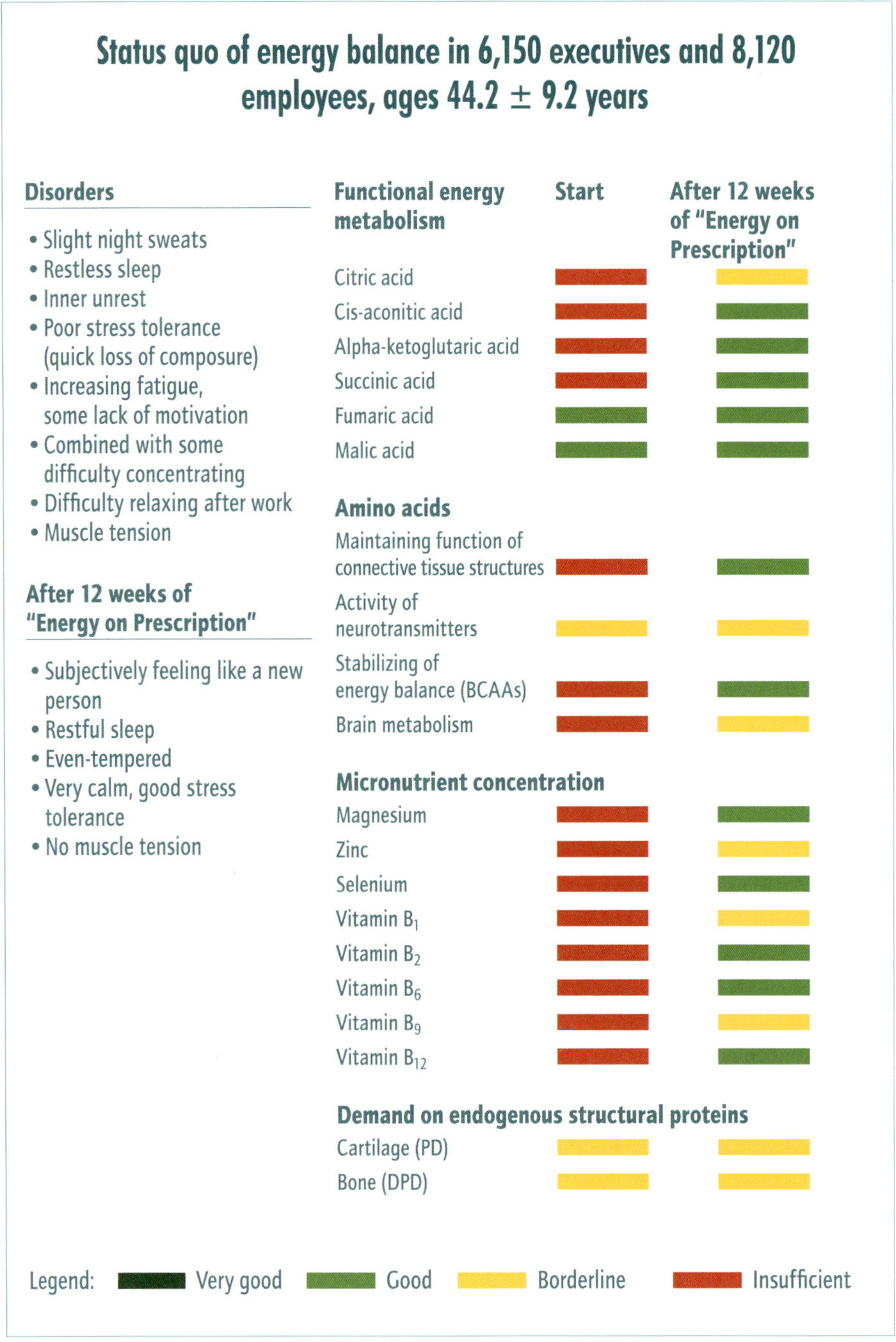

Fig. 6: Status quo of energy balance in 6,150 executives and 8,120 employees ages 44.2 ± 9.2 years

PLEASE NOTE

Our "Energy on Prescription" concept is the customized intake of insufficient micronutrients as a therapeutic dose based on extensive blood tests and urine analyses, a structured case history, and has nothing in common with the arbitrary intake of so-called *nutritional supplements*.

2.2 Wellbeing is measurable

When the energy status of the tested individuals changes, it is always accompanied by positive effects on the individual's sense of wellbeing. A reduced stress index, a reduced vegetative quotient, the LF/HF-ratio, and an increase in pNN50, which is considered the indicator for the parasympathetic nervous system, show the targeted effects of micronutrient therapy on the balance of the vegetative nervous system.

Sense of wellbeing and the balance of the vegetative nervous system can be measured via a simple method, the 24- or 48 hour heartrate variability analysis (HRV). This analysis is an optimal indicator of the vegetative nervous system's balance.

2.2.1 24- or 48-hour heart rate variability analysis (HRV)

Respiration-related fluctuations in the heartrate have been known for centuries. Heartrate variability (HRV) represents the physiological cardiac regulatory mechanisms. The heart must permanently adapt to changing internal and external demands and receives the essential information for doing so from the brain in the form of parasympathetic and sympathetic efferents.

The analysis of HRV is thus a dynamic measuring parameter to ascertain the status of the vegetative nervous system. If the modulations that result from this information are very minor, it is difficult for the heart to adapt to the changed conditions. Functionality is limited.

THE POWER OF MICRONUTRIENTS

During a state of rest, the heart beats an average 60 times per minute, which is equivalent to an average one second or 1000 milliseconds between heartbeats, but the distance actually varies.

The HRV-analysis reveals findings about the balance between the sympathetic and parasympathetic nervous system, about performance readiness, about relaxation ability, as well as sleep and stress load. The chronogram (fig. 7) shows the day profile of the 24-hour HRV-measurement and is a three-dimensional image showing the heartbeat's frequency components.

- High frequency band (HF power) at 0.15 – 0.40 Hz,
- Low frequency band (LF power) at 0.04 – 0.15 Hz and
- general physiological activity of 0.0033 – 0.04 Hz.

The chronogram provides information about the relationship between tension phases and rest phases via its coloring. We use the analysis system "Faros" by the Finnish company Bittium, Kuopio, for the 48-hour HRV-analyses.

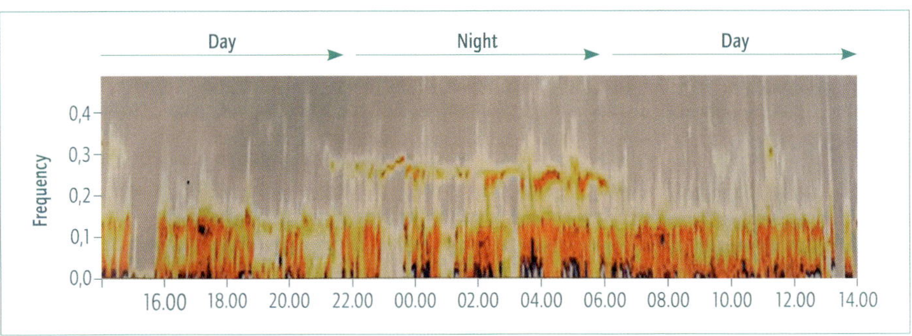

Fig. 7: chronogram

The HRV-analysis determines values for the following parameters: stress index, pNN50, HF power (high frequency), LF power (low frequency), LF/HF-ratio. Test subjects are given a special questionnaire for customized evaluation.

The HRV is based on the optimal synergism of the sympathetic and parasympathetic nervous system (fig. 8). A healthy heart beats irregularly because it constantly tries to create a balance between the sympathetic and parasympathetic nervous system.

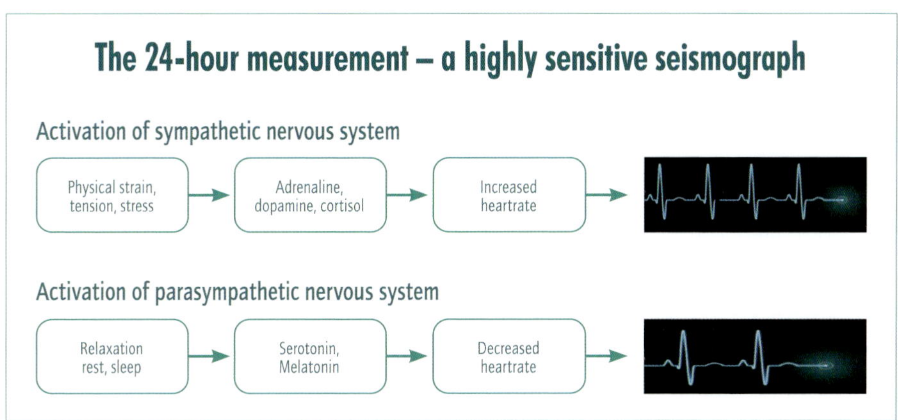

Fig. 8: 24-hour HRV-analysis: optimal parameter for balancing the vegetative nervous system

The 48-hour HRV-analysis can provide important findings, whereby the data is inextricably linked to the patient's case history and activity profile during measuring. The findings include evidence on performance readiness, the ability to relax, sleep architecture, stress load, and the balance between the sympathetic and parasympathetic nervous system (fig. 9).

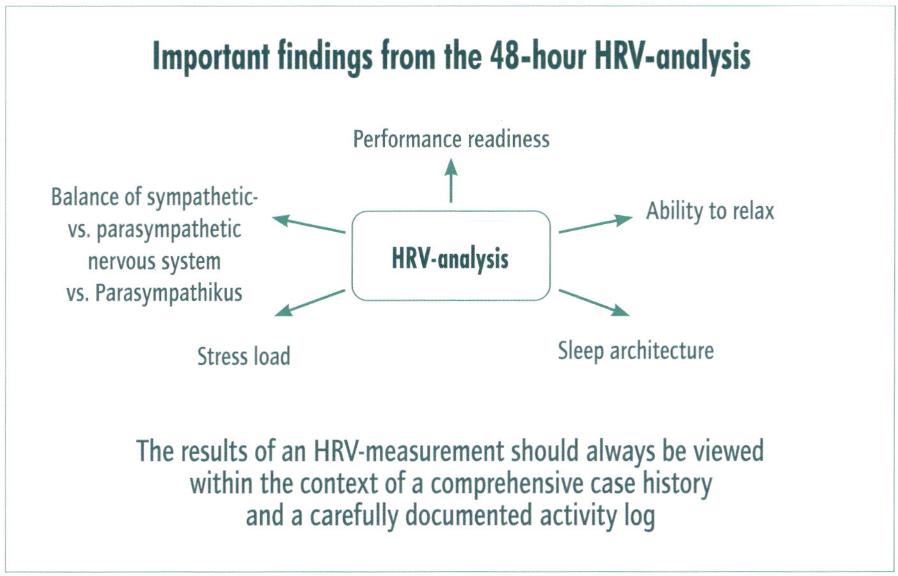

Fig. 9: Important findings that can be gained from a 48-hour HRV-analysis

The individual parameters of the 48-hour measurement are:

Stress index

The *stress index (SI)* is a mathematical quantity with a sensitive reaction to shifts in the vegetative balance:

$$SI = \frac{Amo}{2 \cdot Mo \cdot MxDMn}$$

Amo: Modus amplitude (modus frequency as % of total);

Mo: Modus (most frequent value of RR interval)

MxDMn: Range of variability (maximal RR interval - minimal RR interval)

A low value indicates a low stress load.

pNN50

The parameter pNN50 represents the percentage of neighboring RR-intervals that deviate more than 50 milliseconds from each other in the entire recording.

$$pNN50\ \% = \frac{\text{Number of neighboring intervals with deviations} > 50\ ms}{\text{Number of intervals}}$$

This value is used as a stable indicator for activity of the parasympathetic nervous system. In people younger than 50, this value should be > 10 %. But this value varies with the individual and can verifiably be changed for the better with a targeted intake of micronutrients. A pNN50 > 30 indicates a complete overload of the adrenergic system. Our evidence-based retrospective studies have been able to show this.

HF (High-frequency) power

In a *high-frequency band*, the parasympathetic nervous system is especially active. A high value indicates an active parasympathetic nervous system and thus the recovery phases. By contrast, low values can be an indication of stress.

LF/HF-ratio

The *vegetative quotient* is used as the parameter for balance between the sympathetic and parasympathetic nervous system. High values show a primarily sympathetic nervous system dominance. But to qualify it must be noted that the sympathetic and parasympathetic nervous systems do not always behave reciprocally. Hence this parameter is not beyond dispute. The parameter does not have to correlate with other HRV-analysis parameters. An LF/HF-ratio of < 2 indicates a very good balance of the vegetative nervous system.

2.2.2 Stress hormones in everyday life

Stress hormones include cortisol, a vital steroid hormone in the adrenal cortex that regulates key functions of cellular homeostasis and the immune system. Cortisol can be identified in saliva and reliably represents biologically active (free) cortisol in serum. Impaired function of the adrenal cortex leads, on the one hand, directly to endocrinologic disorders such as Cushing's, and on the other hand to mental disorders, and conditions like lethargy, depressive moods, difficulty concentrating, weight gain, head, muscle, joint, and back pain, and other illnesses.

Cortisol is subject to a 24-hour rhythm (fig. 10) that is coded by the central "inner clock" located in the hypothalamus.

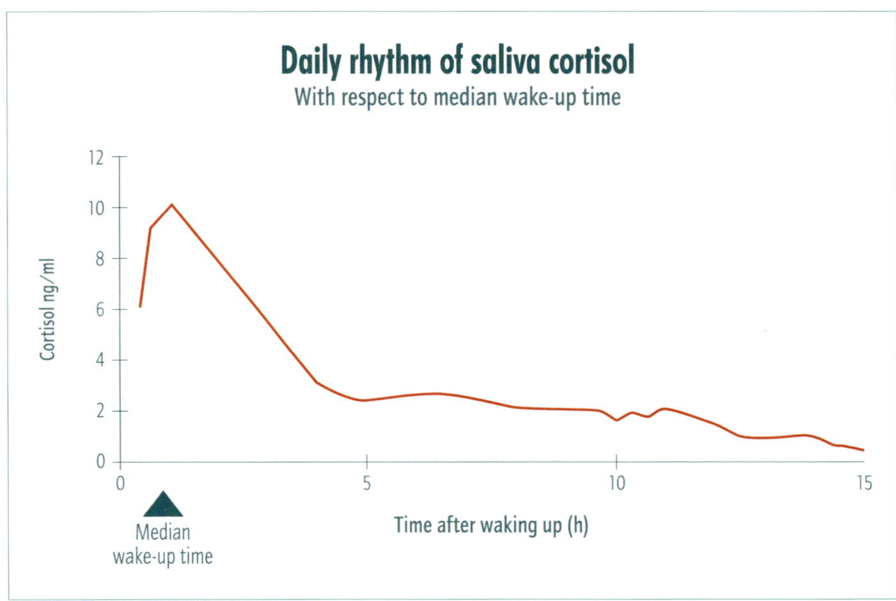

Fig. 10: Cortisol in saliva over the course of the day

- Cortisol starts to build during the night. The maximum amount is available in the morning when waking up.
- After waking up there is a brief additional increase in the cortisol level, the so-called CAR *(Cortisol Awakening Response)* that masks the basic rhythm, followed by a quick drop in the cortisol level.
- The nightly minimum enables anabolic and regenerative activities during the night.
- Cortisol in serum is largely bound to carrier proteins (95-98% bound to CBG or albumen); only a small amount is not bound (1-5%) and is biologically active.
- Unbound cortisol diffuses freely in saliva and reflects the dynamics of adrenaline secretion much better.
- The daily cortisol rate is a direct reflection of the individual stress status.

Morning cortisol values of < 5 ng/ml suggest a strikingly higher stress load, while morning cortisol values of > 8 ng/ml indicate good, balanced sleep behavior and thus good regeneration.

PLEASE NOTE

During analysis it is important to note that results in people doing shift work and those taking certain medications are not considered valid.

Daily cortisol profiles are indicated for people with disorders for whom the 24-/-48-hour measurements cannot provide information about the balance of the vegetative nervous system.

PLEASE NOTE

We refer to hypocortisolism when the adrenergic system shows cortisol values of < 4 ng/ml throughout the day.

A SALUTO research project about stress levels in teachers shows that the stress profile of teachers already shows a severe dysregulation of the vegetative nervous system in the morning (fig. 11):

- A significantly elevated stress load in the morning: < 5 ng/ml. 72% of teachers have values of < 5 ng/ml in the morning. This is an indicator for a very strained vegetative nervous system.
- Morning cortisol values of > 8 ng/ml indicate good, balanced sleep behavior.

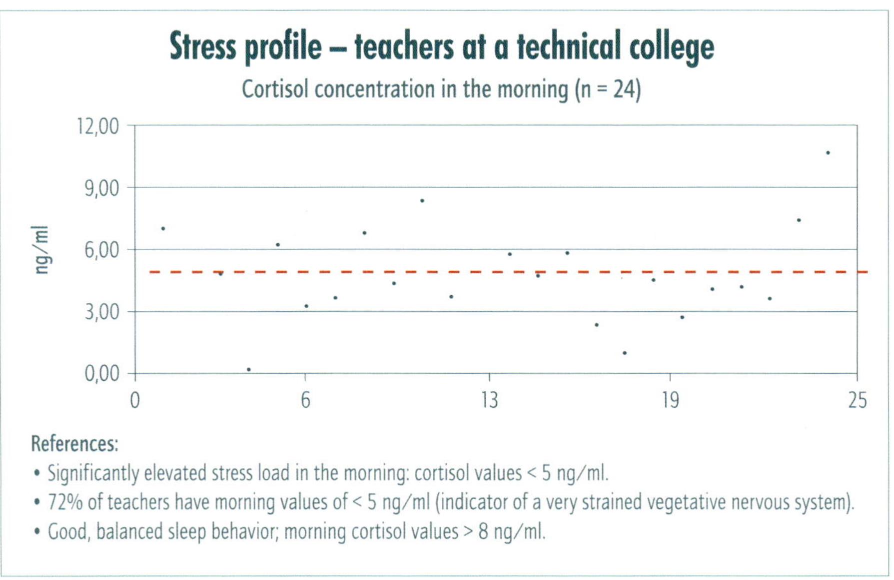

Fig. 11: Morning cortisol values of 24 teachers at a technical college

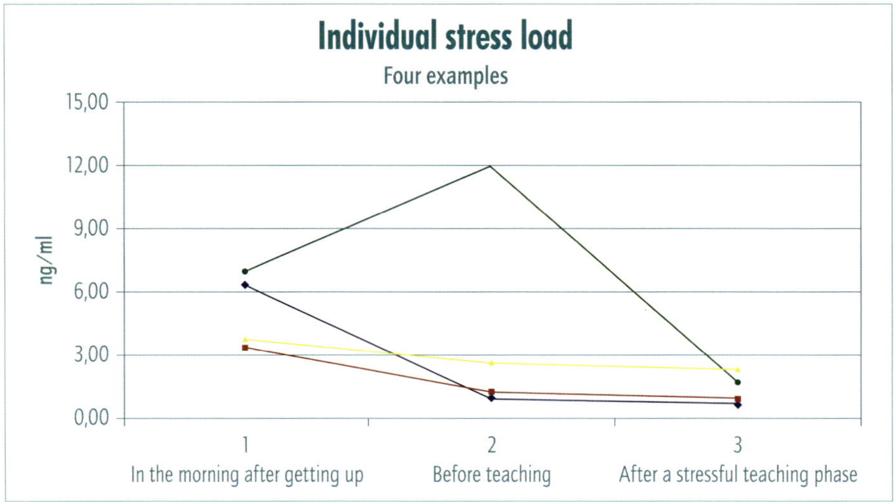

Fig. 12: Four stress profiles: the green curve shows a still functioning adrenergic system; the three remaining curves (red, blue, yellow) show definite signs of absolute chronic fatigue syndrome.

The cortisol day profiles are a good way to show individual stress resistance. When comparing four teachers at a technical college (fig. 12), only one showed a functioning adrenergic system while three colleagues showed signs of chronic fatigue syndrome.

CONCLUSION

The studies and research projects completed by SALUTO show that the customized micronutrient intake verifiably resulted in the normalization of the cortisol day profiles.

2.3 Differentiated energy and micronutrient diagnostics are the Alpha and Omega

Functional energy metabolism

A brief digression on functional energy metabolism (fig. 13; ➡ chapter 6.2.1): The citric acid cycle (synonymous with citrate cycle) is the central switch point for the entire metabolism (fig. 14). It is where the degradation pathways of carbohydrates, fats, and proteins in the form of activated acetic acid (acetyl-CoA) come together. Acetyl-CoA is metabolized for energy production.

Moreover, the citric acid cycle via its intermediates is the starting point for several biosyntheses, e.g. for amino acids glutamine, arginine, proline. An overproduction of intermediates due to a coenzyme and cofactor deficiency, particularly an insufficient supply of coenzyme-Q10 and α-liponic acid, results in mitochondrial dysfunction and limited energy production.

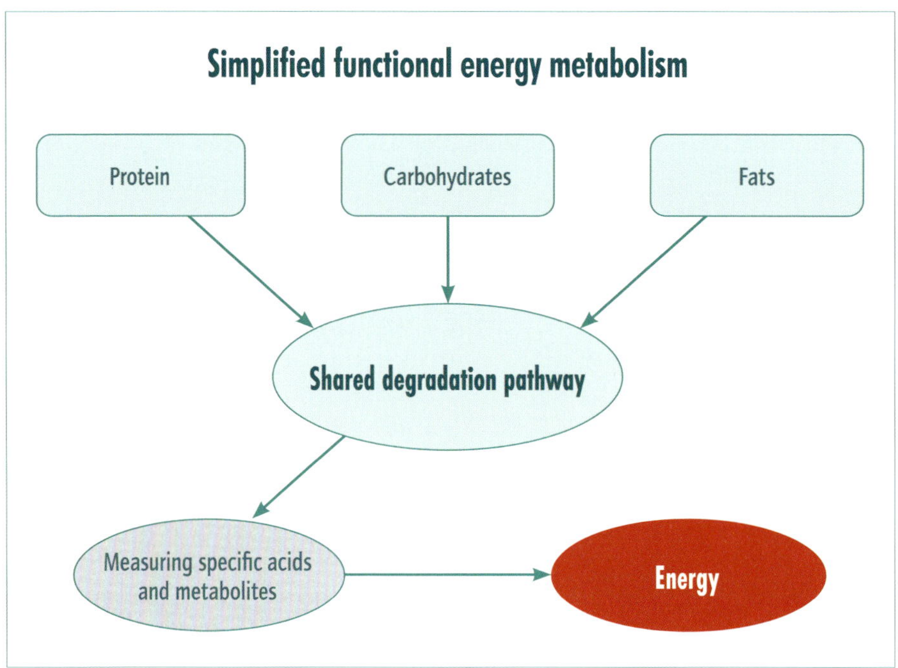

Fig. 13: Simplified functional energy metabolism

Lower or elevated concentrations of metabolites are often caused by an inadequate supply of carbon atoms that are transferred by amino acids to the citric acid cycle's intermediates. The result is a reduction in energy production. A lack of basic micronutrients (amino acids, vitamins, minerals, trace elements) results in a disruption in the "cellular powerplant" (mitochondrial dysfunction).

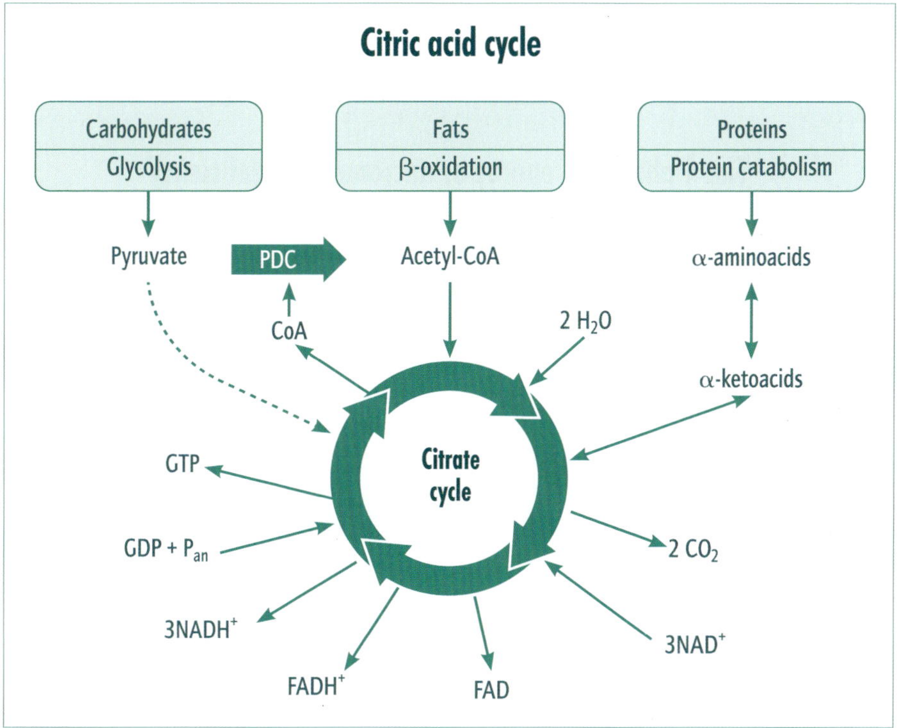

Fig. 14: Citric acid cycle

Various metabolic products are measured within this "energy context" (➡ chapter 6.2.1.1). An elevation and/or reduction in measured substances indicates certain disruptions in the energy metabolism and is linked to a reduction in energy production (fig. 15):

- Phase 1: depletion of stores in tissue and cells.
- Phase 2: increasing exhaustion.
- Phase 3: functional disorders with symptoms that require treatment.
- Phase 4: pathological disorders with irreparable damage.

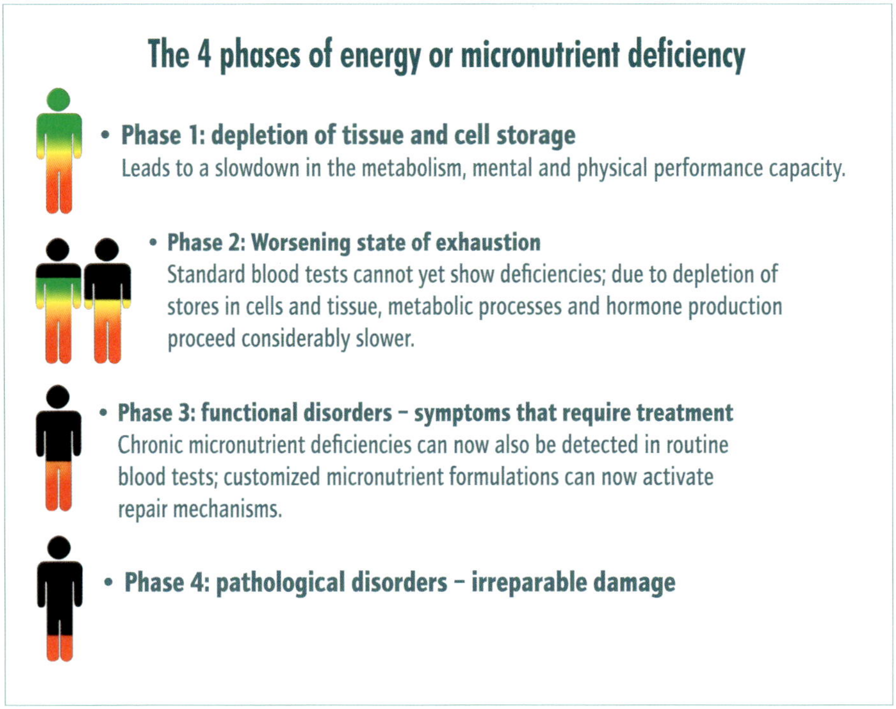

The 4 phases of energy or micronutrient deficiency

- **Phase 1: depletion of tissue and cell storage**
  Leads to a slowdown in the metabolism, mental and physical performance capacity.
- **Phase 2: Worsening state of exhaustion**
  Standard blood tests cannot yet show deficiencies; due to depletion of stores in cells and tissue, metabolic processes and hormone production proceed considerably slower.
- **Phase 3: functional disorders – symptoms that require treatment**
  Chronic micronutrient deficiencies can now also be detected in routine blood tests; customized micronutrient formulations can now activate repair mechanisms.
- **Phase 4: pathological disorders – irreparable damage**

Fig. 15: Four phases of energy or micronutrient deficiency

2.3.1 Micronutrient deficiencies often go undetected

Digression on the components of blood

Blood is the transport medium and thus, in a manner of speaking, the body's liquid transport tissue or rather the circulatory system's helper. Its job is to supply every cell with the vital things, such as fuels from food, oxygen, vitamins, hormones, and warmth, and to eliminate end-products of metabolism and absorb heat from every cell.

The total amount of circulating blood in a human body is approx. 8% of bodyweight. Hence someone weighing 80 kg has an approx. blood volume of 6.4 liters. Blood consists of liquid and solid components: liquid components are referred to as *plasma*, solid components are referred to as *corpuscles*.

Blood plasma is a clear liquid. It consists of 93% water and contains substances such as sodium chloride, carbohydrates, fats, and proteins. Bacteria, viruses, funguses, and parasites that enter the body through skin injuries, respiration, or digestion are fought against. *Red corpuscles* (erythrocytes), with approx. 45% of the total volume (fig. 16), transport oxygen or rather carbon dioxide and contain the red blood pigment (hemoglobin).

During the initial years, all of our blood tests were measured in erythrocytes (intraerythrocytic) as well as serum. Today micronutrient concentrations are measured exclusively in the erythrocytes (intraerythrocytic) because routine serum tests verifiably cannot produce meaningful results.

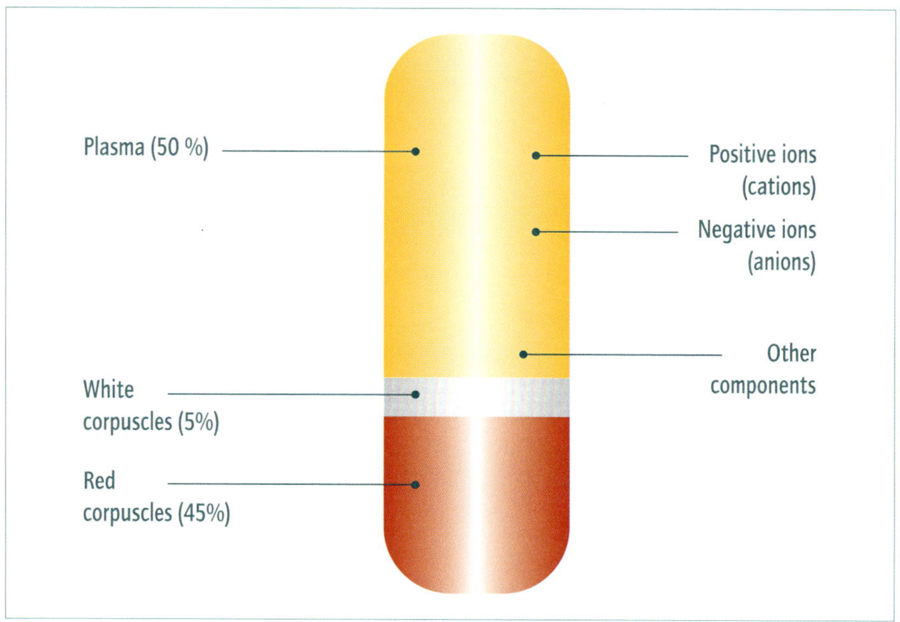

Fig 16: Blood components

Next to liquid blood (plasma), blood contains different blood cells. All of them, with the exception of lymphocytes, which also form in lymphatic organs, form in red bone marrow (medulla ossium rubra). After a certain maturing period, they are released into the blood. The amount present in blood can vary greatly.

Red corpuscles (erythrocytes) have a lifespan of approx. three months, after which they are replaced. A man's body contains approx. five million erythrocytes per cubic millimeter, a woman's body 4.5 million. Their job is primarily the transport of oxygen from the pulmonary alveoli to the organs as well as carbon dioxide transport from the tissue back to the pulmonary alveoli.

If at all, vitamins and trace elements are measured in blood serum but not in the cell since routine blood tests are done at the serum level (fig. 17) where red and white corpuscles are no longer present. Thus, they provide extracellular values (extracellular = present outside the cell; opposite = intracellular).

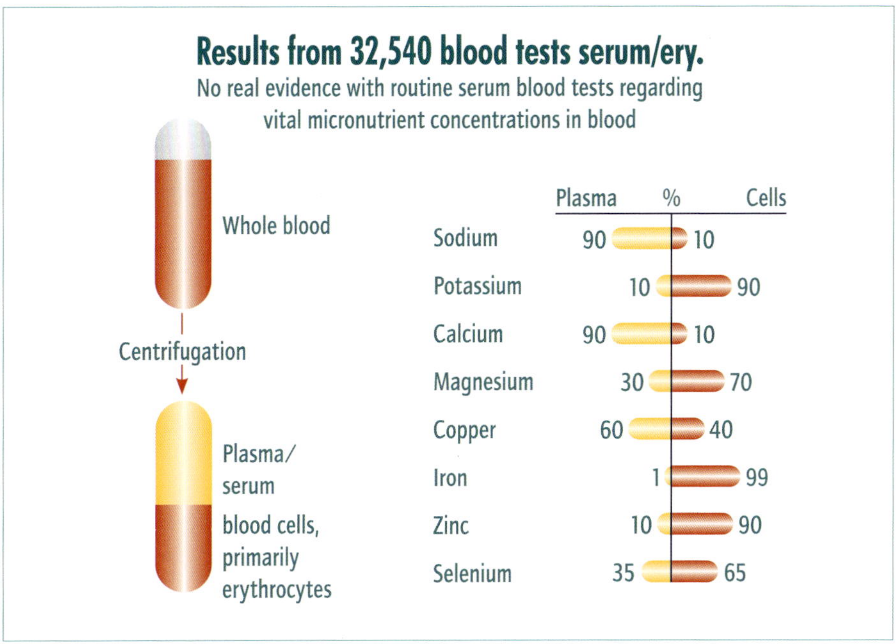

Fig 17: The results from 32,540 blood tests in serum and erythrocytes show that due to serum analyses no real assertions can be made regarding the real values.

With respect to the cell's real micronutrient supply, blood serum measurements are not meaningful (fig. 18) because they do not allow primarily intracellular elements such as potassium, magnesium, zinc, and B-vitamins (B_1, B_2, B_6, B_9) to be captured.

Element determinations in serum do not provide reliable conclusions regarding the mineral, vitamin, and trace element balance. A drop in levels can only be detected when the characteristic deficiency symptoms or even tissue and organ damage appear. But latent undersupplies can have existed for some time without any obvious acute health problem.

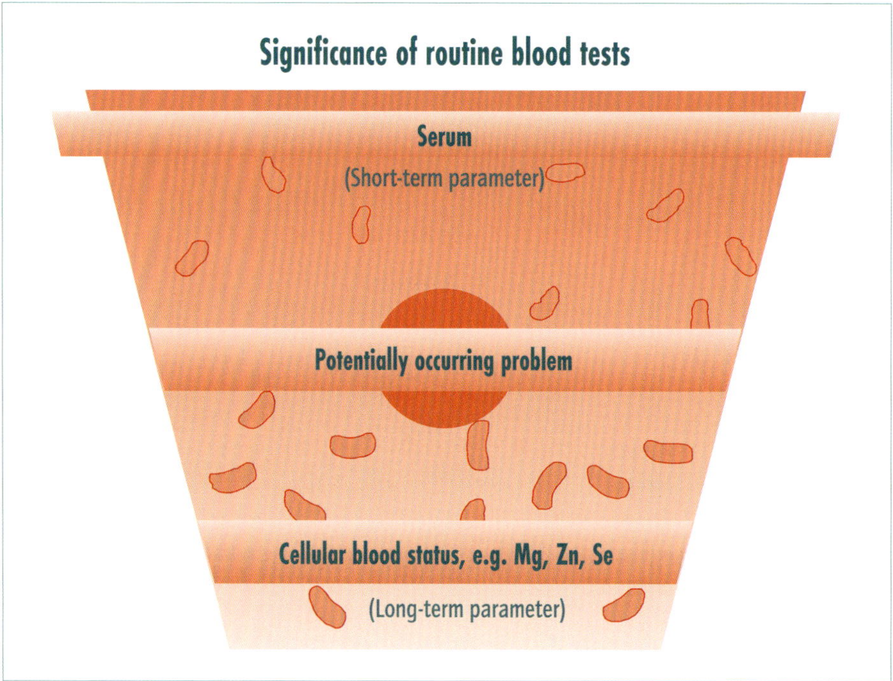

Fig 18: Significance of routine blood tests

The exchange between extra and intracellular space is regulated by complex endogenous mechanisms. The concentration of important elements in serum is kept as constant as possible. If the micronutrient intake (vitamins, minerals, etc.) is not sufficient, endogenous reserves are mobilized to maintain the serum level. That is why most of the time, in-serum deficiencies are detected very late or not at all.

Routine analyses often do not show any deficiencies on the serum level, while definite deficiencies can be detected on the cellular level. Once deficiencies can be detected at the serum level, they have been detectable for a long time on the cellular level.

Disorders can be present for a long time (fig. 15) without an explanation from routine blood tests: decreasing concentration, increasing fatigue, muscle tension, all the way to pollen allergies are only a few of the multi-layered symptoms.

But even when there are no current symptoms, the goal should be not to use up one's personal resources in the first place. We were able to detect inner discontent, poor sleep behavior, decreasing ability to concentrate, and many other disorders in the entrepreneurs, executives, and executive staff we tested. This could be prevented by replenishing cellular stores in a time manner, thereby boosting the functional energy metabolism.

PLEASE NOTE

All possible endogenous resources are mobilized, particularly on the cellular level, to maintain important micronutrients on the serum level. Hence deficiencies are not detectable or detected very late on the serum level.

Optimizing the functional energy metabolism via a targeted intake ("Energy on Prescription") of micronutrients would not be possible without a cellular analysis (➡ chapter 6.2).

PLEASE NOTE

An intraerythrocytic analysis of micronutrients is optimal, since even whole-blood analyses (fig. 19) still show fluctuations in measurements due to a short-term change in diet. But those can be reduced with a fasting blood test administered in the morning.

The whole-blood analysis

Whole-blood analysis is more meaningful since the blood is not centrifuged after it has been collected and contains both intracellular and plasma components.

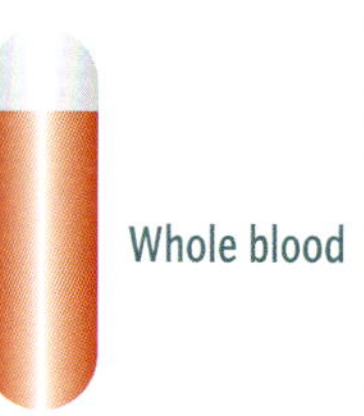

Disadvantage: The blood values are affected by short-term dietary changes.

Fig. 19: Whole-blood analysis

The case reports show that the analyses of intraerythrocytic micronutrient concentrations as long-term parameters are most meaningful. Meanwhile different institutions offer whole-blood analysis, but which, for this reason, has only limited significance because, for instance, magnesium is detected up to 90% intraerythrocytic and not in whole blood. This should absolutely be taken into consideration.

2.3.2 Paradigm shift in laboratory diagnostics via a globally unique database

Past analyses and subsequent evaluations of micronutrient concentrations by the various laboratories were done without a comprehensive basis of individual anamnestic data for individual clients/patients/athletes. How can institutions conduct an evaluation when to some extent only standard values exist for evaluation that are based on the lowest number of test subjects? These are the reasons why in recent years we have built a globally unique database.

The targeted intake of lacking micronutrients and an optimized diet can verifiably help to balance biochemical disturbances. With the help of information scientists, SALUTO has spent years meticulously working on incorporating approx. 60,000 case reports into a computer matrix. The "Individualizer" is a globally unparalleled software that is the basis for the preparation of customized micronutrient formulations (prescription medication) to maintain good health as well as therapy for illnesses without side effects.

The database contains 10,542 target values or rather reference values of micronutrient concentrations, results from years of empirical research projects or rather studies, that were compiled in 297 categories with the respective cluster (gender, age, preexisting conditions, athletic activity) (fig. 20).

Globally unique, evidence-based database opens up new possibilities in micronutrient therapy

59,455 persons tested for specially measured blood and urine values that are differentiated into 10,542 target/reference values divided into 297 categories with the respective cluster:

- Age, gender, mental state, diet, preexisting conditions
- Level of physical activity, type of sport, competitive sports (regeneration, preparation, competition, rehabilitation phase)

The software calculates each person's micronutrient requirement via
- The extensive anamnesis questionnaire
- The measured blood and urine analyses
- The target and reference values in the individual clusters

Fig. 20: Database overview

The software can calculate the individual micronutrient formulations or rather each individual's requirements with the help of the anamnesis questionnaire, results from special blood/urine analyses, and cluster classifications, and can compare them to individuals in the respective age structure, existing disorders or illnesses.

Due to the comprehensive analyses with the help of specially developed algorithms, it is now for the first time possible to determine the micronutrient requirements of individuals and to evaluate them. This database also contains more than 32,500 whole-blood analyses relative to intracellular (intraerythrocytic) blood analyses.

3 ENDOGENOUS REGULATORY SYSTEMS

3 ENDOGENOUS REGULATORY SYSTEMS

3.1 The thyroid – regulates our "sense of wellbeing"

PLEASE NOTE

Vitality and strength with an optimal balance of thyroid hormones!

More than half of the 60,000 tested subjects show various disorders that are largely linked to thyroid hormone regulation. An adequate supply of micronutrients is essential to the complex regulation and synthesis of thyroid hormones. The extraordinary importance of thyroid hormones lies in the division and growth of all cells, the carbohydrate, protein, and fat metabolism, the body's temperature regulation and energy metabolism.

Iodine, Selenium, Vitamin-D, Omega-3 fatty acids, Ubiquinol, and iron control the thyroid's chemical functional sequences and thus the entire energy balance. In the long-term, a micronutrient deficiency can lead to thyroid dysregulation that verifiably can cause various disorders (poor sleep behavior, inner unrest, fluctuating concentration ability, etc.) (Wienecke, 2020).

An inadequate iodine supply can, for instance, cause the thyroid hormone level in the blood to drop. The thyroid reacts with increased growth and enlarges. It tries to balance the lack of thyroid hormones with increased production. A visible sign is the goiter at the neck, also called *struma*.

It is known that disruptions in the production of thyroid hormones affects the organism far beyond the "thyroid" organ. Decades of research has focused on establishing a link between thyroid function and the human body. Latent symptoms are often present – however, often without clinical relevance.

Hormone production control is highly complex and takes place with the help of different hormones and control organs with negative feedback mechanisms (fig 21).

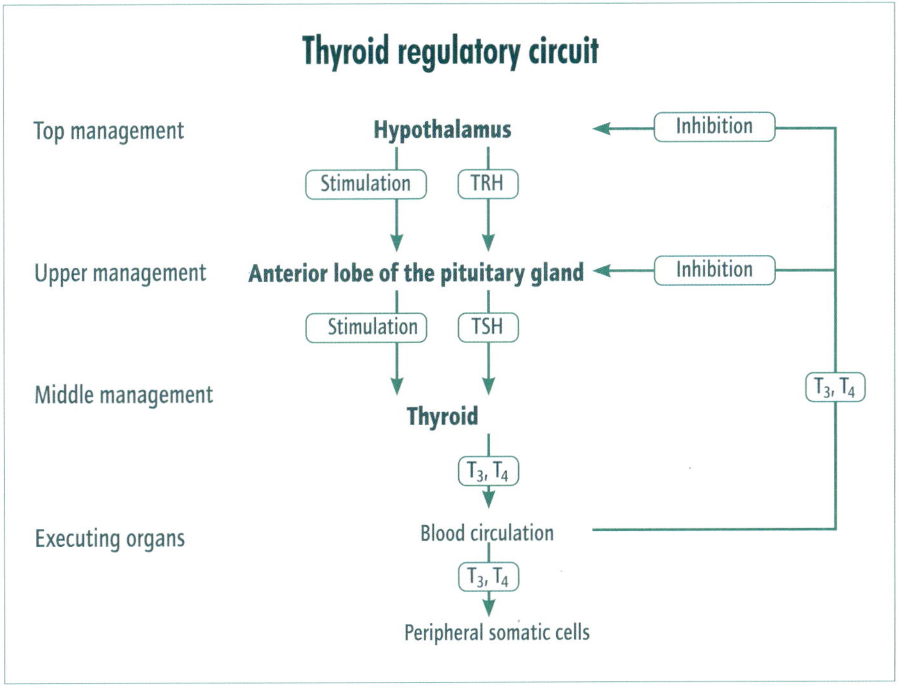

Fig. 21: Control mechanisms in thyroid hormone production T3 and T4. Figure modified as per Wormer, 2014

We will at this point forgo physiological details of thyroid hormones and their control, which can be looked up in the respective professional publications. The object of this content is to express the impact of diet and the targeted intake of lacking micronutrients on thyroid hormone regulation.

Our experiences in recent years show that past reference ranges should be called into question. In the book *Was die Seele essen will* (What the soul wants to eat), the author describes a significantly differentiated view of thyroid hormones and their effects on various disorders (Ross, 2017).

3.1.1 Standard thyroid hormone values

According to medical findings, TSH basal values of 0.4-2.5 µIU/ml are considered "healthy thyroid" (Gronegger, 2015; Zechmann, 2007). Potential autoimmune diseases can be eliminated early with the antibody assessment (thyroid peroxidase, thyrotropin receptor autoantibodies). Next to deficiencies in certain micronutrients such as iodine, selenium, zinc, Omega-3 fatty acids, iron, Vitamin-D, and Ubiquinol, genetic or epigenetic factors can verifiably cause increasing thyroid hormone dysregulation.

Our longtime experience shows that while the effects of thyroid hormones even in the "normal" ranges of 0.4-2.5 µIU/ml are considered as "healthy thyroid", they can in fact cause various disorders.

Latent symptoms have been present for a long time – but so far without clinical relevance

The effects of thyroid hormones on mental and physical performance capacity can be optimally ascertained by measuring the many parameters of the vegetative nervous system (pNN50, stress index, vegetative quotient: LF/HF-ratio; ➡ chapter 2.2) via a 24-hour HRV-measurement. There are currently more than 2,200 scientific publications worldwide that have recognized 24-hour HRV-measurements as an internationally accepted method to ascertain performance readiness, the ability to relax, sleep architecture, stress, and the balance between the sympathetic and parasympathetic nervous systems (fig. 8 and 22).

The following assessment criteria with the different disorders resulted from all of the retrospective study results from more than 5,500 24-hourHRV-analyses – always under consideration of the additionally measured fT_3 and fT_4 values.

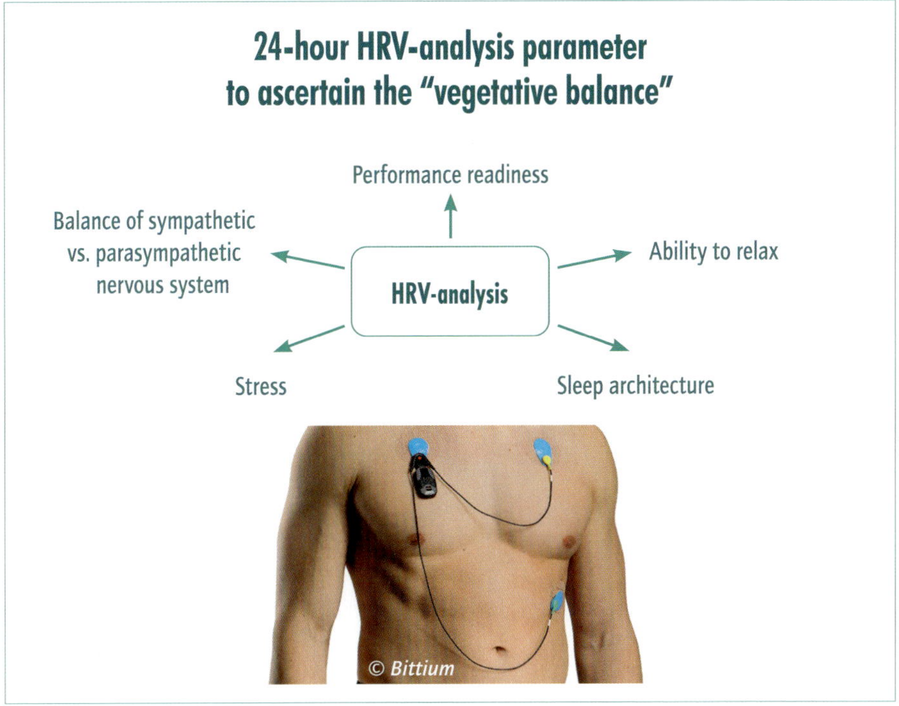

Fig 22: 24-hour HRV-analysis – a parameter to ascertain the "vegetative balance", image rights © Bittium

The 24-hour HRV-measurement as the parameter to ascertain the activity of the vegetative nervous system

PLEASE NOTE

In competitive and elite athletes, thyroid hormones can be higher or lower after intense training sessions and competitions or extreme mental stresses. Extreme physical and/or mental stresses should therefore be avoided on the day before a measurement is taken.

Evidence-based test results from 789 competitive and elite athletes show the effects of thyroid hormones on the vegetative nervous system's various parameters (pNN50, stress index, vegetative quotient: LF/HF-ratio). The extensive 24-hour HRV-measurements have shown that optimal performance potentials can be achieved primarily in the 1.6-2.2 µIU/ml range of the basal TSH-value (fig. 23). However, only 17% (N = 457) of the TSH-values measured in 789 elite athletes were in this range (Wienecke, 2020).

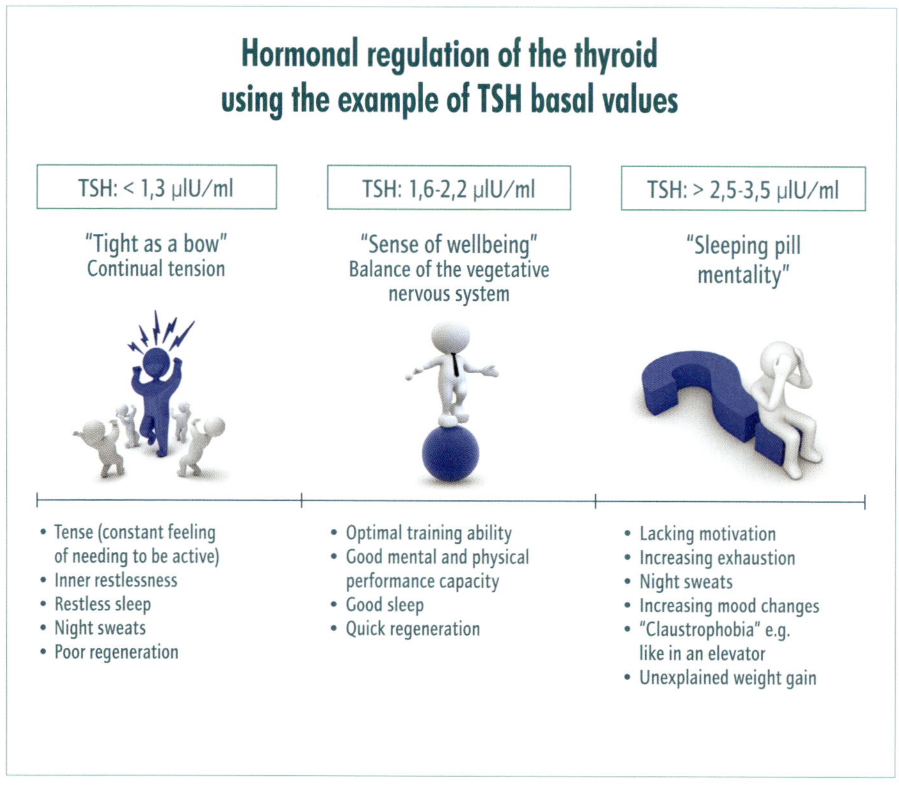

Fig. 23: Hormonal regulation of the thyroid using the example of the TSH basal value

A balance and the optimal regulation of thyroid hormones can be achieved with the aid of the 24-hour HRV-measurement (➡ chapter 2.2), Dr. Mosetter's Glycoplan (➡ chapter 4.7), and the customized intake of lacking micronutrients (➡ chapter 6.3) (fig 23). 25% of elite athletes with a TSH basal value of <1.3 µUI/ml exhibit the disorders listed in fig. 23. 58% of elite athletes with TSH basal values of > 2.5 µUI/ml describe the disorders listed in fig. 23 such as increasing fatigue, etc.

3.1.2 Thyroid hormones and the vegetative nervous system

The table below shows the effect of thyroid hormones, here the TSH basal values, in elite athletes, executives, and "ordinary" people as well as patients, on some of the vegetative nervous system's parameters, whereby the results were ascertained with the help of the 24-hour HRV-measurement.

Comments: Also ascertained at the same time were fT_3, fT_4, and thyroid antibodies (thyroid peroxidase, thyrotropin receptor antibodies). The blood, urine, and HRV measurements were all completed after a non-training day.

Tab. 1: Correlation between parameters of the 24-hour HRV-analysis in elite athletes (total N = 789; pro soccer players N = 125, marathon runners N = 144, other athletes N = 520 such as team handball, Judo, track & field) and the TSH basal values

	TSH basal values N = 134 (Gr. 1) 1,6-2,2 µIU/ml (1,95 ± 0,31)	TSH basal values N= 197 (Gr. 2) 0,5-1,3 µIU/ml (0,86 ± 0,39)	TSH basal values N = 457 (Gr. 3) 2,5-3,5 µIU/ml (2,91 ± 0,49)	r`° betw. Gr. 1-Gr. 2	r°` betw. Gr. 1-Gr. 3
Stress Index	168,3 ± 47,4	415,2 ± 61,9	445,9 ± 78,1	0,91 ***	0,86 ***
Vegetative quotient LF/HF-ratio	1,72 ± 0,19	7,31 ± 1,51	6,99 ± 1,73	0,87 ***	0,81 ***
pNN50 %	14,98 ± 1,23	5,31 ± 1,51	8,13 ± 1,42	0,79 ***	0,92 ***
r = highly significant correlation factor (p < 0.001) of the TSH basal value and the listed parameters of the 24-hor measurement in elite athletes					

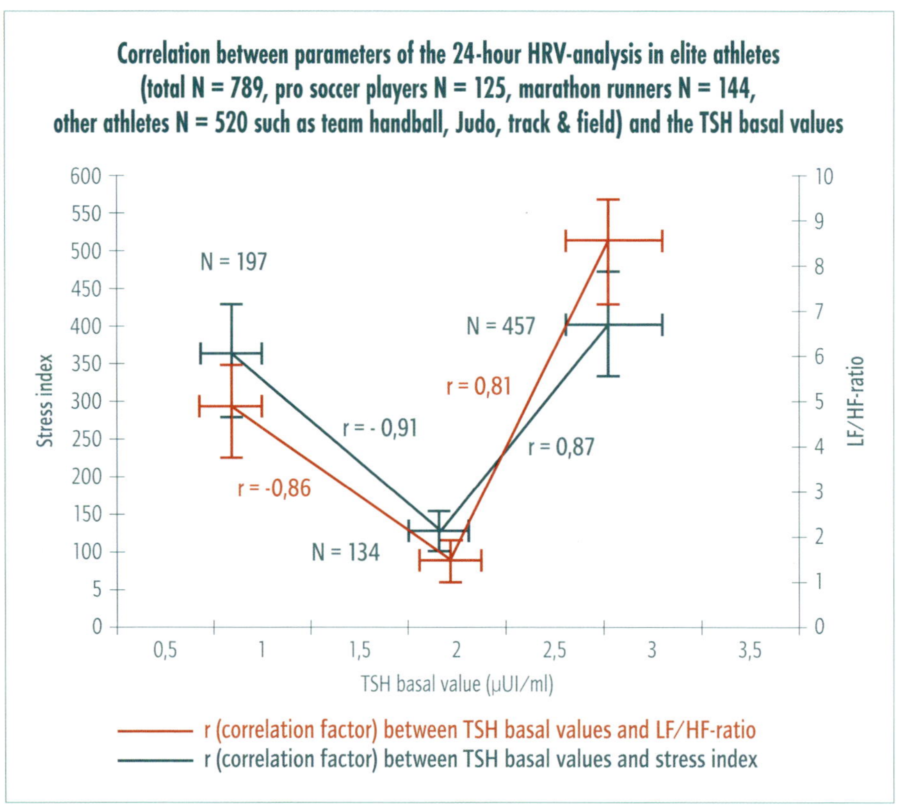

Fig. 24: Correlation between parameters of the 24-hour HRV-analysis in elite athletes (total N = 789; pro soccer players N = 125, marathon runners N = 144, other athletes N – 520 such as team handball, Judo, track & field) and the TSH basal values

Visible are high correlation factors between the different TSH basal values and the parameters for the 24-hour HRV-measurement (tab. 1 and fig. 24). The 197 elite athletes whose basal TSH-values corresponded to the arithmetic mean 0.86 ± 0.39 µUI/ml (tab. 1) showed a high stress index at 415 ± 61.9. When comparing this stress index in the arithmetic mean with that of the 134 elite athletes who are within the arithmetic mean at 1.95 ± 0.31, we can see a highly significant correlation of r = 0.91 (fig. 24; ➡ chapter 6.3.8).

This means that athletes with a TSH basal value < 1.3 µUI/ml with a high stress index already present with the disorders of the "tight as a bow" mentality (fig. 23). The results also show vegetative quotients (LF/HF-ratio) in the arithmetic mean of 5.31 ± 1.51 in

197 elite athletes with a TSH basal value of 0.86 ± 0.39, while the 134 elite athletes with a TSH basal value of 1.95 ± 0.31 showed a statistically highly significant low LF-HF quotient ($p < 0.001$).

As a vegetative quotient, this parameter documents primary sympathetic dominance and can lead to verifiably worse regeneration in the elite athletes after intense training and competition stress. The highly significant correlation factor of $r = 0.81$ (tab. 1 and fig. 24) convincingly shows this connection.

Elite athletes with TSH basal values between 2.5 and 3.5 µUI/ml show the listed disorders of "sleeping pill mentality" (fig. 23). The elite athletes in this reference range also show a high stress index (tab. 1 and fig. 24) and with 6.99 ± 1.73 a high vegetative quotient (LF/HF-ratio).

For the stress index the highly significant correlation factor between the TSH basal values lies at $r = 0.91$ (tab.1 and fig. 24) and for the vegetative quotient (LF/HF-ratio) at $r = 0.87$.

These are significant criteria for training management. Normalization of thyroid hormones in these athletes can be achieved with a targeted dose of iodine and selenium. If that is not sufficient, a dose of, for instance, L-thyroxin 50, would be beneficial. The desired target range for the TSH basal value: 1.6-2.2 µUI/ml.

CONCLUSION

Of the 789 elite athletes, only 17% (N = 134) were within the desirable (Wienecke, 2020) reference range of 1.6-2.2 µUI/ml in which, based on our experience, trainability is optimal. 25% (N = 197) were in a reference range of 0.5-1.3 µUI/ml with the described "tight as a bow" mentality, and 58% (N = 457) were in a reference range of 2.5-3.5 µUI/ml with the described "sleeping pill" mentality.

These results show an urgent need for action. The targeted intake of insufficient micronutrients can verifiably improve the described disorders, but in some cases can also normalize an existing "dysregulation" of thyroid hormones.

Additional findings of the evidence-based retrospective study results with the 24-hour HRV-measurement, a customized micronutrient intake, and a change in diet (Glycoplan as per Dr. Mosetter) show new, indicatory therapeutic approaches for "ordinary people", competitive athletes, and patients overall.

The customized micronutrient intake for, e.g., 162 ADHD patients (➡ chapter 6.3.1) after a successful intervention over a period of 24 weeks, verifies that the pNN50 as the indicator for the parasympathetic nervous system in the reference range between 1.5-2.5 µUI/ml shows a mean value of 15.1 ± 1.07 and presents an optimal balance of the vegetative nervous system, while the test subjects with a TSH basal value < 1.5 µUI/ml and >2.5 µUI/ml show a markedly reduced pNN50.

The same is true for the stress index and the vegetative quotients (LF/HF-ratio). The lowest stress index shows up in the range of 1.5-2.5 µUI/ml with 181.44 ± 32.85. The vegetative quotient (LF/HF-ratio) in the range of 1.5-25 µUI/ml shows an optimal value of 1.64 ± 0.44.

3.1.3 Dysregulation of thyroid hormones forces malaise, anxiety, and depressive mood

A study by Siegmann et al, completed in 2018, was able to establish a link between thyroid hormone level, depression and anxiety. An analysis of several studies done with a total of 44,388 participants (patients with autoimmune thyroiditis, comparable thyroid disorders versus healthy test subjects) showed that patients with autoimmune thyroiditis were 3.3 x more at risk of developing depression. The risk of developing anxiety was 2.3 x higher.

Experts estimate that approx. 45% of depression cases and 30% of anxiety cases are accompanied by autoimmune thyroiditis. This link between mental disorders and thyroid function was awarded so much significance that on May 4, 2018, J. Köppe wrote in *Spiegel online* that depression and anxiety are common, as are thyroid disorders, and that doctors have now shown that the two often coincide, and that this finding could help affected patients.

We were able to show that a targeted micronutrient intake can normalize or regulate thyroid hormones and thereby recharge vital energy.

A targeted micronutrient intake can normalize or regulate thyroid hormones

250 individuals ages 33.4 ± 5.4

First group: 84 test subjects before supplementation:	TSH basal values: 0.89 ± 0.23 μUI/ml
Second group: 104 test subjects before supplementation:	TSH basal values: 3.41 ± 0.39 μUI/ml
Third group: 62 test subjects before supplementation:	TSH basal values: 1.91 ± 0.29 μUI/ml

Thyroid antibodies TPO/TRAb show no signs of an autoimmune disease

After a customized micronutrient intake

Test subjects in the first group show:	TSH basal values: 1.68 ± 0.19 μUI/ml
Test subjects in the second group show:	TSH basal values: 1.98 ± 0.22 μUI/ml
Test subjects in the third group show:	TSH basal values: 1.87 ± 0.18 μUI/ml

Fig. 25: A targeted micronutrient intake normalizes or regulates thyroid hormones

Our studies show: In 250 individuals ages 33.4 ± 5.4, whose thyroid hormone levels were outside the desirable reference range of between 1.6 and 2.2 μUI/ml, the targeted intake of insufficient micronutrients was able to verifiably achieve normalization (fig. 25). The measurements, taken with the help of 24-hour HRV-measurements, show a markedly improved stress tolerance and the activation of the parasympathetic nervous system (relaxation of the vegetative nervous system).

CONCLUSION

Normalizing thyroid hormones with insufficient micronutrients results in a well-balanced inner "sense of wellbeing".

3.2 "The gut – the other brain"

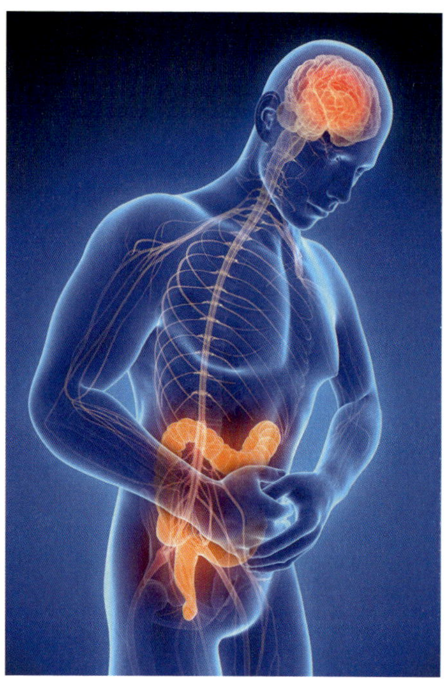

Fig. 26: The gut – the other brain

In recent years, the human microbiome has taken on increasing importance. Every human body is host to an estimated 40-100 billion microorganisms or microbes, including bacteria, viruses, and funguses. Most live in the digestive tract, others on the skin, in skin creases, and on mucosa. Collectively the genes of these living organisms are called *microbiome*. All of the organisms combined are also referred to as *microbiota*. Each body hosts about as many microbes as it has cells.

Bacteria communicate with the brain

We do not yet know the exact mechanisms with which the gut and the brain affect each other (fig. 26). Researchers are pursuing various hypotheses: Cells in the intestinal wall produce most of the peripheral, meaning outside the brain, available serotonin, an important neurotransmitter. The peripheral serotonin could possibly affect the serotonin level and thus signal processing in the brain.

The microbiome also interacts with the immune system and regulates, for instance, cytokine activity. These proteins participate in inflammation processes in the brain; hence there is a possible link between microorganisms and nervous affections like multiple sclerosis. In addition, microbial metabolic products, so-called metabolites, seem to change the structure of the blood-cerebral barrier.

The gut-brain axis

1. Gut flora affects behavior and mood by way of bacteria and the brain communicating via the bloodstream and the nervous system.
2. Some mental disorders and developmental disorders affecting brain function can be linked to a deviation in the composition of these microbiota. This is suspected to be the case with certain forms of autism, depression, and anxiety disorders.
3. A series of studies, to date mostly done with animals, show promise that some of these illnesses and defects can be mitigated by normalizing the microbial balance in the gut.

For this reason, deliberate nutritional behavior as outlined in Dr. Kurt Mosetter's Glycoplan (➡ chapter 4.7) and the basic principles of clean eating, are of fundamental importance for a successful micronutrient therapy.

Every analysis also includes measuring of the intestinal fatty acid binding protein (I-FABP) values to check for existing gut barrier disorders.

I-FABP parameters to check for gut barrier dysfunction

The intestinal fatty acid binding protein (I-FABP) is currently the best-suited marker for an impaired gut barrier. I-FABP occurs exclusively in the cytoplasm of intestinal epithelial cells (100% specific!) and plays a role in the fatty acid metabolism. When the intestinal epithelium is damaged, I-FABP is released into the circulation and can be measured in serum.

Compared to zonulin, this parameter has more informational value.

If the I-FABP values are elevated harmonization of the gut flora is recommended before starting micronutrient therapy. This measurement can rule out a leaky gut.

Leaky gut:

- Leaky gut is caused by chronic inflammation when the barrier function of the small intestine's mucous membrane is impaired.
- This allows bacteria and toxins from the intestinal mucosa to enter the bloodstream and promotes systematic inflammation.
- Increased permeability of the intestinal mucosa is also associated with other diseases (e.g. diabetes mellitus, rheumatoid arthritis).
- Characteristic of a leaky gut is the increased permeability of the "tight junctions" located in the intestinal mucosa or functional or structural damage to the intestinal epithelial cells.

3.3 The effects of "electrosmog" on our "sense of wellbeing" and the vegetative nervous system

The effects of electromagnetic radiation on our "sense of wellbeing", which the author analyzed on his own person, documents a strong activating effect on the balance of the vegetative nervous system and sleep behavior. With the help of the 24-hour HRV analysis we are able to show that cellphone use during the day with and without suppressor in the form of Gabriel Technology (chip and bracelet) has verifiably led to increased sympathetic nervous system activity.

The 24-hour HRV-analysis parameters make it possible to show especially at night how the Gabriel Technology has resulted in a marked reduction and thus "damping" of the vegetative nervous system. In spite of the author's very good preconditions, after the long-term intake of a customized micronutrient formulation, the parameters of the 48-hour HRV-measurement did show additional slight improvements.

Overall, the stress index dropped from 250 to 162 (fig. 27). The pNN50 (fig. 28) as indicator for the parasympathetic nervous system rose from already good 7.83 to 11.2, the vegetative quotient LF/HF-ratio (fig. 29) dropped from 3.51 to 2.12 and documents the positive effects on the balance of the vegetative nervous system.

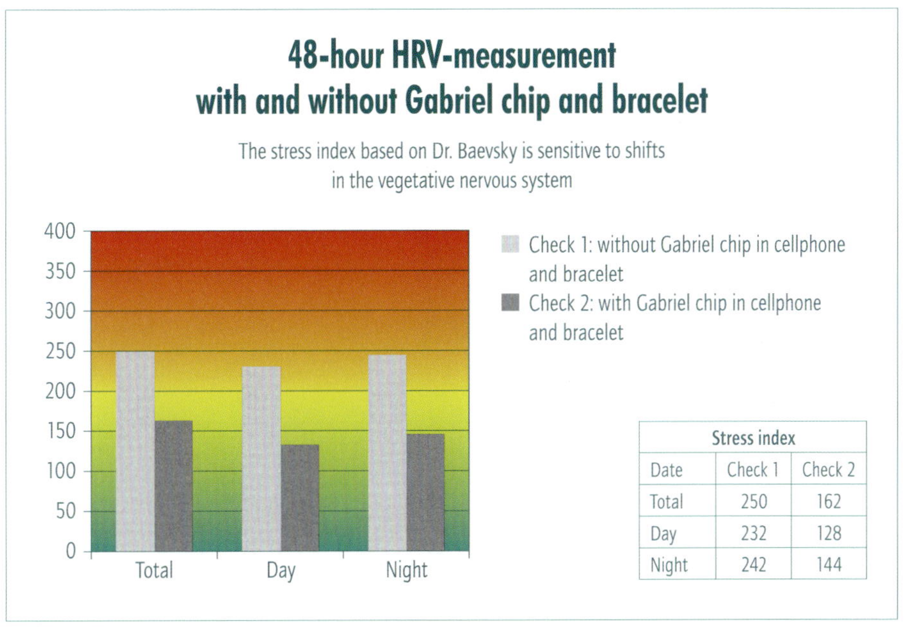

Stress index		
Date	Check 1	Check 2
Total	250	162
Day	232	128
Night	242	144

Fig. 27: Stress index in the 48-hour HRV-measurement with and without Gabriel chip and bracelet

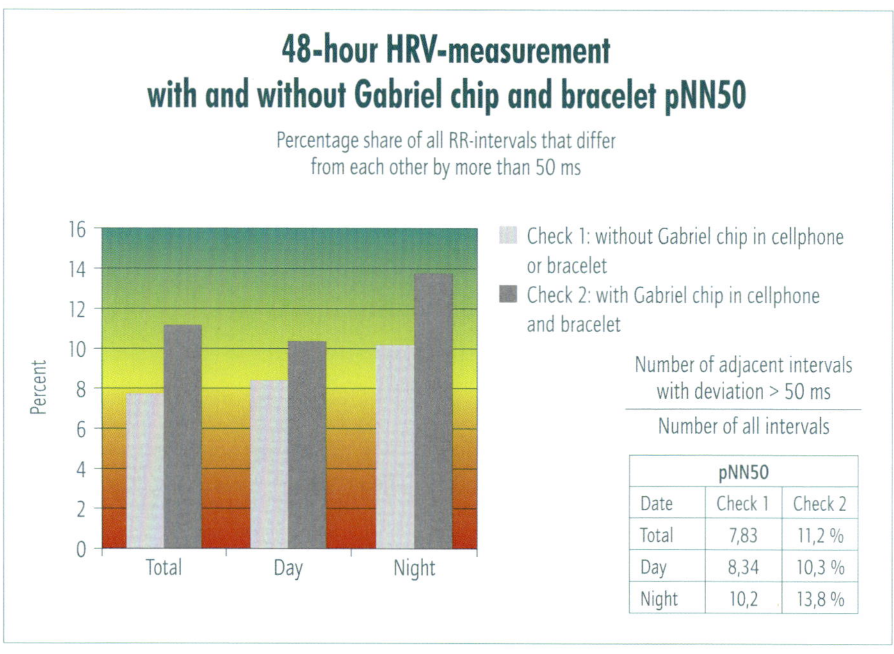

pNN50		
Date	Check 1	Check 2
Total	7,83	11,2 %
Day	8,34	10,3 %
Night	10,2	13,8 %

Fig. 28: pNN50 in the 48-hour HRV-measurement with and without Gabriel chip and bracelet

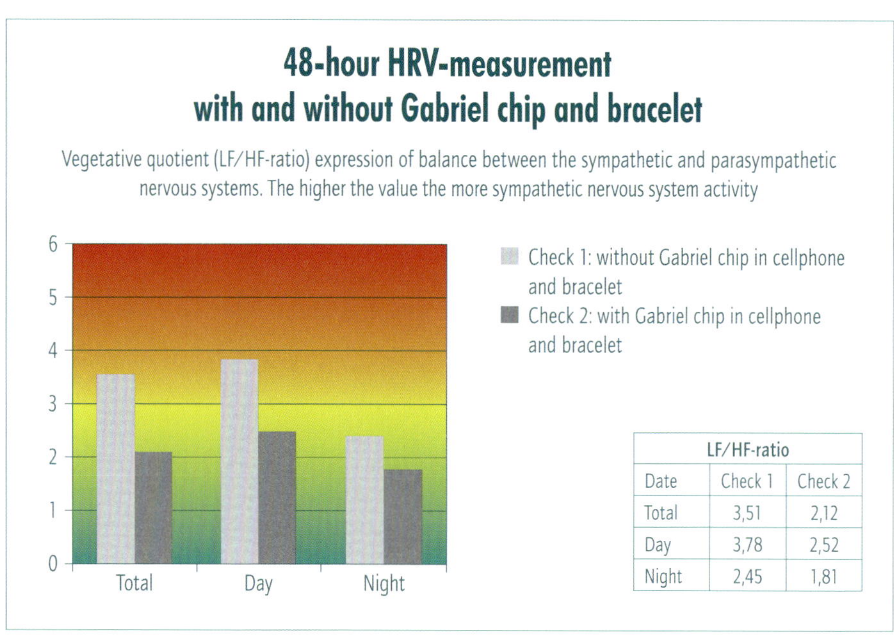

LF/HF-ratio		
Date	Check 1	Check 2
Total	3,51	2,12
Day	3,78	2,52
Night	2,45	1,81

Fig. 29: LF/HF-ratio in the 48-hour HRV-measurement with and without Gabriel chip and bracelet

The author's subjective sense of wellbeing during his self-test is confirmed by the result from the 48-hour HRV-analysis. Analogously additional results from 123 test subjects show the positive effects of the Gabriel technology on the vegetative nervous system balance.

But there are now more studies on the effects of electromagnetic radiation on the brain metabolism. Protection from the effects of electromagnetic radiation on the brain's activity and metabolism is a significant element of regeneration.

Mainz University tested five cellphone chips from different manufacturers at the behest of the SfGU (Foundation for Health and the Environment). While stress activation in the brain was significantly reduced with the use of the Gabriel chips, no effect could be detected with the other tested chips (fig. 30).

The joint pioneering achievement of SfGU and Gabriel Technology received scientific validation from a neutral source in 2018, when the effectiveness of the Gabriel chips was made public in the professional publication *Frontiers in Neuroscience* (Lausanne/ Switzerland) (Henz et al., 2018). Electric brain activity was recorded by 128 electrodes, which were attached to the surface of the head according to the international 12-20 system, via a so-called *high-density electroencephalogram* (EEG).

The activation sources in the brain were localized with the help of a 3D-analysis of brain activity. The EEG case-by-case testing was conducted with five different cellphone chips. In addition, each test subject also underwent a control measurement without chip application. Fig. 30 shows the radiation exposure from the different electronic devices.

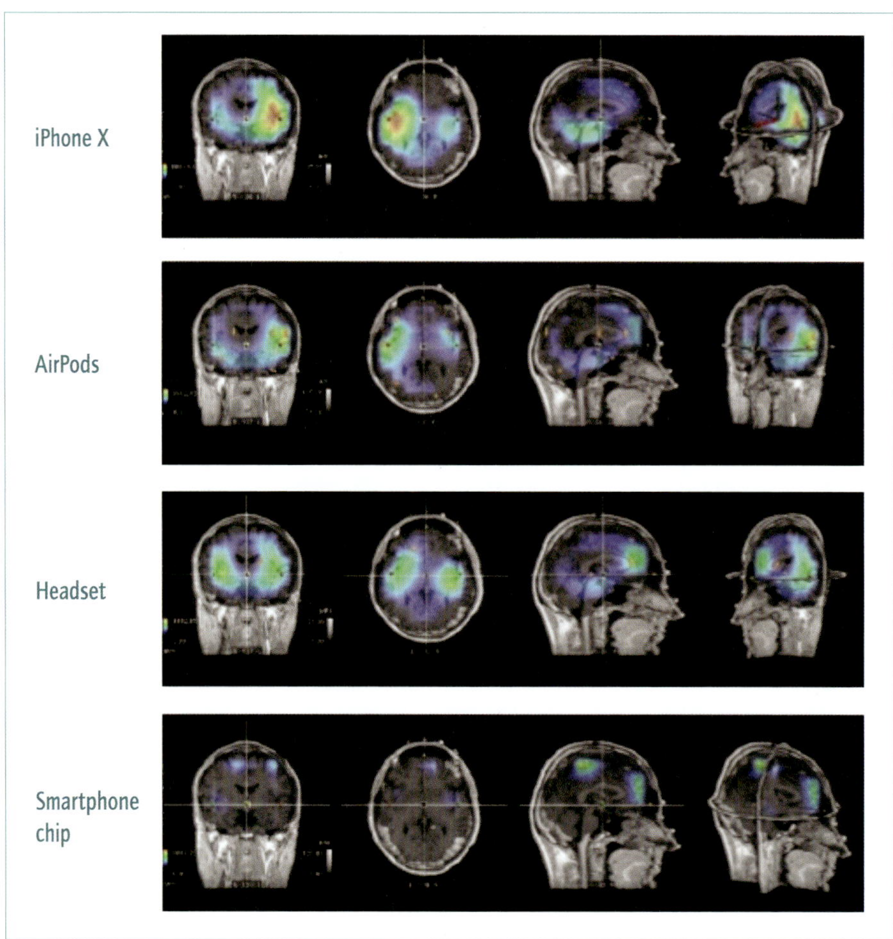

Fig. 30: Effects of a 10-minute radiation exposure from the iPhone X, Bluetooth AirPods, and cable headsets on brain activity

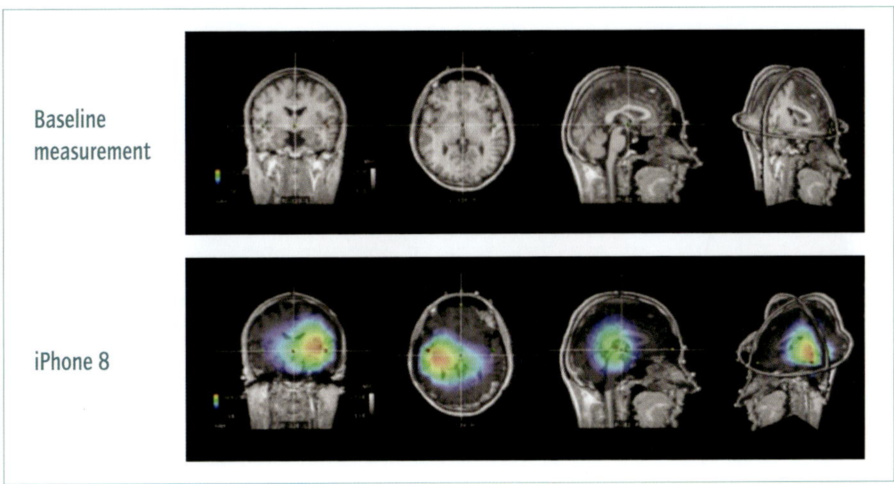

Fig. 31: Effects of a 10-minute exposure to radiation from the iPhone X, Bluetooth AirPods, and cable headsets on brain activity. Strong activation could be detected within the beta and gamma band in the temporal brain areas and parts of the limbic system, the Hippocampus (Henz, 2019)

Electromagnetic radiation or activity from mobile communications, among others, affects certain areas of the brain metabolism. The radiation exposure causes a change in the natural composition of frequency bands theta, alpha, beta, and gamma activity (fig. 32). These frequency bands show the brain's fundamental functional statuses. Mobile communications activate primarily the frequency bands alpha, beta, and gamma.

This affects mental, cognitive functions and with frequent use can lead to increased activity of the sympathetic nervous system, delaying, for instance regenerative processes in athletes, among others replenishment of the glycogen stores in the liver and muscles after previous training and competition loads.

Initial studies also show reduced reaction and concentration ability (so far unpublished). However, controversial discussions in science about this are ongoing. It took three years just to find an applicable journal to publish the Mainz University study (Henz et al., 2018).

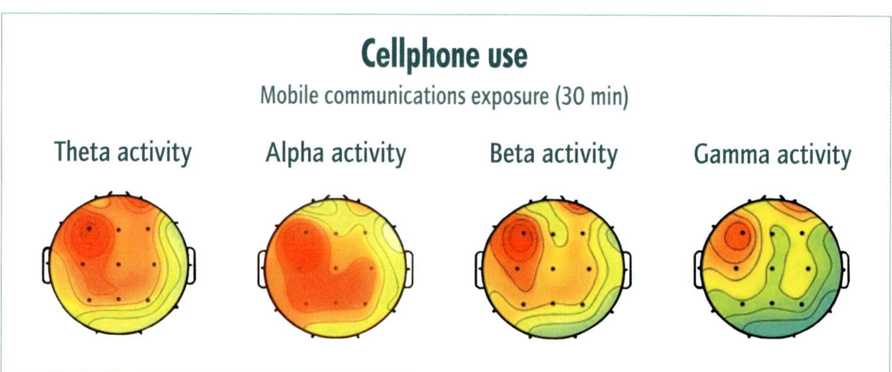

Fig. 32: Effects of 30-minute cellphone exposure on brain activity (Henz, 2019)

Summary of the study Frontiers of Neuroscience (Henz, et al., 2018):

"The results of this study largely coincide with those from earlier studies in the area of neuroscience that tested the effects of EMS exposure on brain activity from the use of cellphones. The increased EEG activations were achieved in all tested frequency bands by exposure to electromagnetic fields from cellphones. A decrease in the EMF-induced activations is observed when using a cellphone chip, particularly in the areas of high-frequency (beta and gamma wave frequency bands). This is observed both at a resting state and while working. A more in-depth analysis of EEG signals shows that while using the cellphone chip fewer points of origin of brain activation are found when the brain is exposed to EMF's emitted by cellphones compared to test conditions where a placebo chip or no chip is used. The results of this study encourage further studies on the long-term effects of the use of cellphone chips in mobile phones in the work environment, as they have a positive effect on brain activity with the short-term use of cellphone chips."

3.4 Exercise promotes endogenous regulation

3.4.1 Our real treasure chest – achieving mental strength and joy of living through "mindful movement"

Mindful movement (mindful walking, powerwalking, and running) is meditation in motion. Mindfulness and body-awareness exercises are combined before, during, and after running. This unique exercise promotes physical and mental fitness. As a combination of mindfulness and moderate exercise, mindful movement promotes stress reduction, improves performance capacity, and is conducive to a happy and fulfilled life.

Mindful movement is suitable for

- Anyone that enjoys moderate exercise while enjoying nature;
- and wishes to improve their performance long-term without suffering from exhaustion.

What does mindful movement do for you?

- Stress relief, physical and mental relaxation;
- Inner peace and tranquility;
- Improved body image
- Being more in the "here and now", less brooding;
- More joy, lightness, and variety while running;
- Awareness in managing one's own resources;
- Fewer injuries;
- Improved performance via mental strength and positive thought management;
- More energy through mindful movement.

THE POWER OF MICRONUTRIENTS

Overall interest in health topics is greater than ever before. Ten years ago, many of these people visited fitness studios. But all too often that membership was the result of a heroic decision without much consequence. Signing up and working out regularly are two very different things. And thus, it happened that many of the supposedly active exercisers were actually studio or club members on paper only. Why should I exhaust myself in my free time and look bad doing so, just to be completely wiped out in the end? Sore muscles guaranteed!

Our experiences in recent years show that people with extremely demanding jobs look for a similarly intense exercise experience. Their motto is "more is more" – far from it. Their maxim is to utilize that limited amount of time as intensively as possible.

Too much is not healthy

The ladder to success must be climbed one rung at a time. Anyone who thinks he will be able to achieve top athletic performances in the limited amount of time he has at his disposal, is sorely mistaken. It is not possible to run three marathons a year and create an optimal healthy balance. Even in sports, the dosage determines whether physical exercise promotes health or weakens it and makes someone vulnerable. There is plenty of proof for how excessive ambition can be damaging.

People who exercise based on the motto "higher, faster, farther" are seriously mistaken. Here, too, it is important to balance exertion and recovery. "Energy on Prescription" is one of the mainstays of mental and physical performance capacity.

Better radio communication in the immune system with mindful movement

Our findings show that people who regularly practice mindful movement have a considerably lower risk of infection. The body's immune defense forms an extremely complex system that responds to outside influences with a number of reactions. Moderate athletic activity improves the quality of highly competent immune cells, so-called *killer cells.*

But the activation of myokines in a very fit individual can also be detected by a high density of binding sites on the surface of these cells that attract viruses and pathogens, but also tumor cells, more effectively and destroy them. This effect of an increased sensitivity of immune cells is primarily due to stimulation by certain messengers, the interleukin, whose production increases after athletic activity.

"Internal communication" of the immune system is boosted by muscular reactions and our body's immune defense is put in a more active and "vigilant" state.

Mindful movement activates myokines

Myokines are hormonelike endogenous substances which the muscle releases directly with increased muscle activity. From a biochemical perspective, they are part of the interleukin group, which in turn is part of the peptide hormones. The name was coined by Danish professor Bente Klarlund Petersen from Righospitalet at the University of Copenhagen in 2007. Myokines are considered messengers with a highly diverse positive effect on the entire organism.

Mindful movement inspires calm and confidence – the best stress killer

This is true at least for moderate physical activity, because everyone knows that exhausting exercise is not an effective stress killer. When you exert yourself excessively and frequently, certain stress hormones are released, especially adrenaline. You might feel well subjectively because your vegetative nervous system is "amped up", but you will not find real inner peace and relaxation.

Excessive exercise in the long-term even has a damaging effect on your health status. Conditions such as inner restlessness, pent up aggression and tension can be relieved through mindful movement, for instance with a daily moderate 20-minute endurance workout. No medication has a more positive impact on your personal condition – and it does so with absolutely no damaging side effects.

3.4.2 The best brain exercise ("Life Kinetik"): Continuously new movement patterns

Life Kinetik uses the body to deploy the reserves of the brain. Improving performance by tapping into reserves in the structural, biochemical, and mental sphere to create the best conditions for daily thought processes and tasks in everyday life and in sports. Since this type of training involves little physical strain, it is appropriate for everyone's personal development regardless of age and fitness level.

Just one hour per week is sufficient to notice the first changes after only a short period of time. Many former and active elite athletes, skiers, biathletes, soccer players, team handball and tennis players, currently benefit from doing this training once a week.

Fig. 33: Areas of cross sections of the brain of a 62-year-old woman before and after three weeks of daily 3-minute coordination training

A human being is born with 200 billion neurons. The best way to link these neuronal structures is by performing specific movements that can activate different areas of the brain simultaneously. Moderate exercise and specific, simple coordination exercises use very different areas of the brain. The objective of these exercises: Linking and creating information highways that, regardless of age, still lead to positive changes in neuronal structures.

Figure 33 was produced as part of a student research paper we created in collaboration with a special imaging method used by radiologists. It is quite impressive to see that after just 5 minutes a day of special coordination exercises additional areas of the brain are utilized after only three weeks, thereby achieving improvement in neuronal structures long-term.

Our practical experiences showed that people who regularly perform this type of movement training significantly improve their mental farsightedness at work through conscious perception, are able to connect the dots more quickly during meetings, and are better able to follow visuals and be more attentive.

But competitive and elite athletes also benefit from these varied movement tasks. They reduce expenditure of energy and effort, achieve elegant and harmonious execution of difficult movement sequences, and improve spatial perception and orientation.

CONCLUSION

Daily 5-minute coordination training with elements from Life Kinetik can result in significantly altered cross-linking of neuronal structures.

3.4.3 Measuring exertion – your pulse rate

"Running without panting" used to be the motto for a healthy way of exercising. But it's not quite that simple, because depending on the type of sport and the temperature, you can overexert yourself and not realize it in time. Checking the pulse rate is therefore essential to engaging in well-balanced exercise. Someone who burns approx. 2000 kcal a week by engaging in athletic activity is practicing optimal fitness and exercise. A man weighing 90 kg completing a 30-minute endurance run is equivalent to approx. 400 kcal.

How to test your capacity

You can determine the ideal heart rate for your exercise with this simple test. While it is not as exact as a professional stress test in a specialized facility, it is helpful when getting started. All you need is a heart rate monitor. This is how it works:

Determining the ideal heart rate:

- Use the following formula to structure a test protocol:
  Resting heart rate .... BPM
  Heart rate during 4 x 4 breathing rhythm ...BPM
  Heart rate during 3 x 3 breathing rhythm ...BPM
- Measure your resting heart rate: Sit down and relax for 3 minutes, then check your pulse with the heart rate monitor and record the result in the test protocol.
- Take four steps while inhaling once; take another four steps while exhaling once. Practice this 4 x 4 breathing rhythm for about 5 minutes.
- Now increase your speed for as long as you are able to maintain the 4 x 4 rhythm. After walking for 10 minutes, check your heart rate and record the result in the test protocol. The pulse rate you have ascertained this way is the target rate you should aim for in the future while exercising.
- Now increase your walking/running speed so you inhale/exhale once for every three steps you take. With this 3 x 3 rhythm you will reach an exercise intensity that an unfit person is initially unable to maintain for even just five minutes. After about five minutes of walking or running, record your heart rate at this level of intensity.
- Exercising in a 2 x 2 rhythm should be avoided: It results in additional activation of the vegetative nervous system and increases the stress in people who are already stressed in their private and professional lives.

This is how you exercise properly

Once you have completed the stress test you can start your personal exercise program.

This is how it works:

- Choose the 4 x 4 rhythm while walking and aim for the target heart rate ascertained with your stress test. If you get winded you are going too fast. The motto is: success without panting. This allows you to avoid feeling exhausted and will greatly improve your performance capacity, stress tolerance, and quality of life within just eight weeks.
- While previously inactive people will begin by walking, physically active people will most likely have to start out running to reach their target rate. But the pace and the distance covered are completely irrelevant. What matters is the amount of time you exercise. Begin by exercising three times a week for 30-40 minutes each (e.g. Monday, Wednesday, Friday). Later you can exercise every day if you like.
- The level of your ideal heart rate also depends on the type of exercise, so you can't simply apply your ascertained target heart rate to other types of exercise. For cycling and cross-country skiing, the heart rates are about 8-10% lower, for swimming, due to the diving reflex effect, it is approx. 6-7% lower, than for running.
- After just a few weeks of exercising you will no longer be able to reach your target rate with the same pace. So now you have to increase your pace to be able to aim for your "old" pulse rate. It is best to track your progress via the training protocol.
- In the protocol you can also record your resting heart rate, which you should measure before each training session and which can be subject to fluctuations, for instance due to a minor infection, insufficient hydration, smoking cigarettes, overheating in summer temperatures or when dressed too warm, but also when exercising at higher altitudes in the mountains.

4 NUTRITIONAL MEDICINE

4 NUTRITIONAL MEDICINE

4.1 The brain is what we eat

TIP

The film *Unser Hirn ist, was es isst* (Our Brain Is What It Eats) offers a good overview of this topic (https://www.arte.tv/de/videos/082725-000-A/unser-hirn-ist-was-es-isst/; last access July 21, 2020).

Dr. Felice Jacka, a renowned scientist and director of "Food and Mood Centres" in Melbourne (Australien; https://foodandmoodcentre.com.au/2016/07/diet-in-pregnancy/) studied the effects of diet on the developing brain in more than 20,000 pregnant women. Children whose mothers ate fast food during pregnancy showed a considerably higher "aggression potential" or hyperactivity.

These findings were verified by additional studies done in Norway, Spain, and Canada: They show a strong correlation between a fast-food diet and mood fluctuations as well as autism spectrum disorders in children (https://www.arte.tv/de/videos/082725-000-A/unser-hirn-ist-was-es-isst/; last access July 21, 2020).

Moreover, a fast-food diet damages spatial memory. Dr. Margaret Morris of Sydney (Australien; https://www.arte.tv/de/videos/082725-000-A/unser-hirn-ist-was-es-isst/; letzter Zugriff 21.07.2020) was able to show changes in the hippocampus of test subjects who had been eating a fast-food diet for four days. Cytokine-mediated "silent inflammation" impacts neuronal structures: immune cells eat up glial cells and thus also destroy the neuronal structures (fig. 34). These reactions by the glial cells were filmed in vitro. It is suspected that this phenomenon seriously damages brain function.

TIP

Starting at minute 29:05 of the film shown on *Arte*, *Unser Hirn ist was es isst*, we can see how vigorously these cells go about their work. The fragments of neurons they are eating are marked in red. The activity level of a fat mouse increases greatly. It is suspected that this phenomenon seriously damages brain function (https://www.arte.tv/de/videos/082725-000-A/unser-hirn-ist-was-es-isst/; last access July 21, 2020).

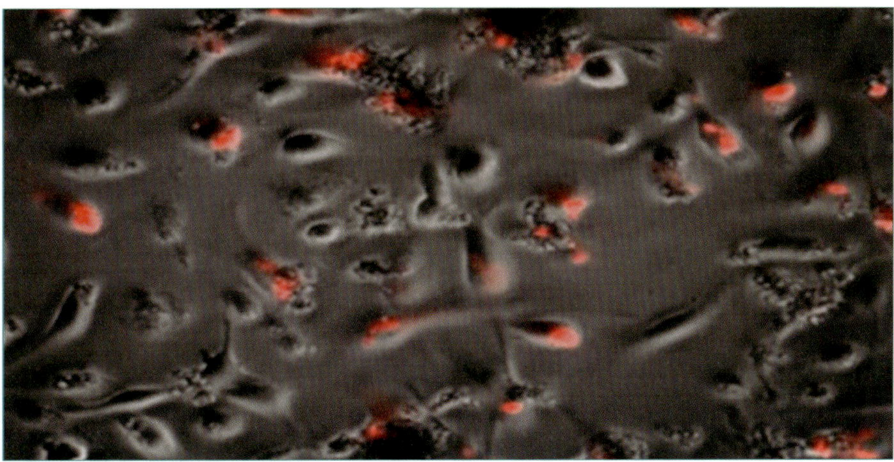

Fig. 34: Immune cells eat up glial cells and destroy neuronal structures with frequent fast-food consumption (results from experiments in mice).

The cause of these effects from a fast-food diet could be a lack of Omega-3 fatty acids, without which the brain cannot develop. 90% of brain matter consists of gray cells that are fundamentally influenced by Omega-3 fatty acids. A good supply verifiably causes improved elasticity of cell membranes and thus better cell function.

Omega-3 fatty acids that reach the brain improve the electric properties of cell membranes that absorb them. Neurons that are rich in Omega-3 fatty acids forward the signal more quickly and functionality is more efficient.

A lack of Omega-3 fatty acids can impact the way the brain functions (fig. 35). The tests we have done with measuring of the HS-Omega-3 index show serious deficiencies. This is true for, e.g. mothers during pregnancy but also while nursing, and thus corresponds to the results Dr. Jacka was able to ascertain. The brain's cognitive performance is crucially impacted by a good supply of Omega-3 fatty acids. Changes in the neuronal structure during Omega-3 fatty-acid deficiency can possibly also lead to communication disorders.

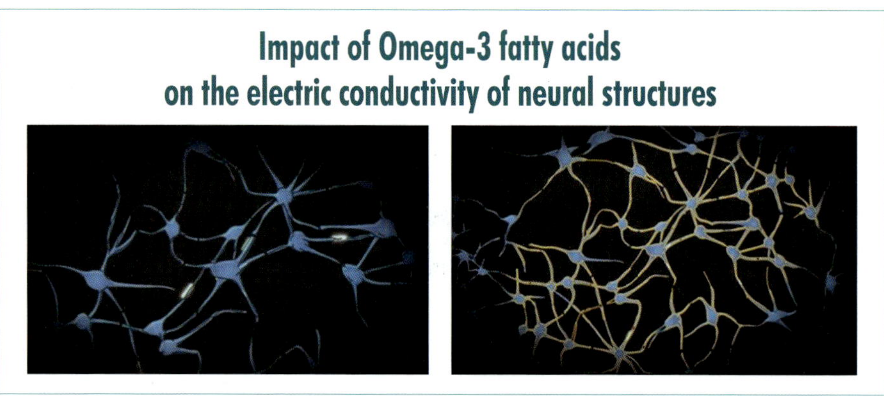

Fig. 35: Impact of Omega-3 fatty acids on the electric conductivity of neuronal structures: before (left) and after (right) intake of Omega-3 fatty acids (From https://www.arte.tv/de/videos/082725-000-A/unser-hirn-ist-was-es-isst/)

TIP

You can find information about Omega-3 fatty acids starting at minute 7:35 of the film "Unser Hirn ist was es isst" shown on *Arte* (https://www.arte.tv/de/videos/082725-000-A/unser-hirn-ist-was-es-isst/; last access July 21, 2020).

Prof. Dr. von Schacky and his team have created a unique process via the HS-Omega-3 index that makes it possible to gather scientifically valid data. The analysis done inside erythrocytes that was established in Europe now shows the many positive effects of an intake of Omega-3 fatty acids for the prevention of and therapy for different illnesses in more than 200 publications. The critical factor for achieving a therapeutic effect is primarily the correct dosage for intake.

The controlled and retrospective study on the customized intake of micronutrients including Omega-3 fatty acids (3,400 mg EPA/DHA in liquid form in two single doses morning and afternoon) of 160 ADHD patients (➡ chapter 6.3.1) verifiably raised the HS-Omega-3 index from 3.51% FS to 8.72% FS while the control group, which merely had a change in diet, showed no change in the HS-Omega-3 index.

Children with ADHD between the ages of 7-12 who did micronutrient therapy showed a significantly improved stress tolerance and ability to concentrate (➡ chapter 6.3.1).

However, not only Omega-3 fatty acids have an effect on the brain, but the brain is also what we eat. The impact of proteins on decision-making ability and sugar on brain function in general has been shown multiple times:

In a study done at Lübeck University (Dept. of Psychology, Dr. So Young Park; Strang et al., 2017), 24 test subjects ate identical-looking breakfasts containing varying amounts of protein. Computer-based tests showed that the higher protein content resulted in improved decision-making ability. The test subjects with a high protein content were more "resistant" to unfair offers, while the reactions of test subjects with a higher carbohydrate content were more sensitive. Test subjects with high tyrosine values were able to increase dopamine release via an improved biosynthesis performance.

At the Oregon Research Institute in the United States, Dr. Eric Stice was able to use MRT to show the effects of ice cream on the body's reward system in 100 students: Students that had not consumed ice cream showed strong activation of the reward system. Students that had consumed ice cream daily for one week, showed no activation of the reward system – the dose was too low, a typical habituation effect. This supports the theory that sugar has addiction potential. Research at the University of Bordeaux (France) showed that glucose increases action potential and changes structures in the brain.

TIP

Information on the sugar consumption experiment can be found starting at minute 35:58 of the film "Unser Hirn ist was es isst" shown on *Arte*, (https://www.arte.tv/de/videos/082725-000-A/unser-hirn-ist-was-es-isst/; last access July 21, 2020).

The Dutch psychologist A. P. Zaalberg researched a link between impulsivity, aggression, and the frequency of legal offenses and was able to show an impact of diet on the potential for violence. He was able to verify this link between diet and criminality in a study: The regular prison food of 221 prisoners who volunteered to participate in the study was supplemented with vitamins, minerals, and amino acids for three months. Questionnaires completed by the participating prisoners themselves as well as guards during the study were analyzed.

Zaalberg et al. (2010) showed that the prisoners' aggressive behavior could be decreased through diet optimization. The number of days spent in solitary confinement could be reduced by one third. Stress tolerance also improved in this context.

In the animal kingdom it was verified that cannibalism in hamsters is caused by a Vitamin B_3 deficiency.

4.2 Fairytales and myths in nutritional medicine

It is true that each year nearly 11 million people die due to an unhealthy diet such as consuming extremely sugary beverages and eating too few vegetables – nearly one in five deaths. These deaths result as long-term effects of a diet that is not adjusted for very long periods of time.

The antiquated opinions of the DGE (German Nutrition Society) can be read nearly daily on social media platforms: In Germany it is very rare for healthy people with a varied diet to suffer from an undersupply of vitamins and minerals.

By 2030, more than 50% of the population will be older than age 50. All of these people have different disorders or illnesses and already take 4-5 medications. The latest studies from around the world show that only 20% of the population has the kind of diet nutrition experts would like. Clean eating is the maxim of nutrition in the future. The Glycoplan as per Dr. Mosetter addresses all of these aspects (➡ chapter 4.7).

Functional energy-metabolism disorders due to micronutrient deficiencies in many medical conditions

Full of energy, fit, and productive for as long as possible, that's what people want. Everyone certainly knows about the close relationship between diet, brain metabolism, immune system and performance. Some may wonder why they don't feel good or aren't productive in spite of a balanced diet.

According to the DGE's latest nutrition report and statements from renowned nutritionists, sports scientists, and sports medicine specialists an additional intake of micronutrients is unnecessary and should be dismissed as long as the individual has a balanced diet (except for fluoride, iodine, and folic acid; here the insufficient intake from a normal diet in Germany is sufficiently well-known).

Our experience in working with more than 60,000 tested individuals, be it executives, athletes, or people with various preexisting conditions, has been quite different. The current energy and micronutrient demands of individuals can only be analyzed via a comprehensive analysis of the functional energy metabolism, intracellular micronutrients, amino acids, and some other basic parameters.

4.3 An engine cannot run without fuel

An adequate supply of micronutrients and trace elements is imperative to maintaining vital functions. Particularly essential is the participation of zinc, selenium, magnesium and all of the B-vitamins (B_1, B_2, B_6, B_9, B_{12}) in physiological processes of cellular metabolism, cell division, stability, energy metabolism, and immune response.

Based on our recent findings, in the long-term a micronutrient deficiency in people with special professional responsibility will lead to many negative health effects in the form of increased chronic fatigue syndrome with cognitive impairments. Competitive athletes significantly increase their injury risk due to premature consumption of the body's own structural proteins that represents a crucial basis for training and competition continuity.

Many disorders in humans with different illnesses and medication verifiably benefit from a targeted intake of lacking micronutrients and an adjusted diet. Our results show that the functional energy metabolism shows restricted activity of certain enzymes that can be ascribed to cellular micronutrient deficiencies in nearly all areas and have caused many disorders. Those affected most are people that are exposed to increasing stressors.

In recent years, some studies (clinical studies) and projects document that even with a balanced diet (based on the criteria of the DGE (German Nutrition Society) a significant micronutrient deficiency in the cell can be ascertained that cannot be adequately corrected without a targeted intake of micronutrients. Here we must point in particular to specific cellular measuring and diagnostics of these deficiencies. Europe has very few laboratory centers that perform these comprehensive functional energy-metabolism analyses and cellular micronutrient analyses and possess an adequate database.

4.4 For many people a "healthy diet" is an unachievable optimum

Everyone suspects that there is a link between diet and performance capacity. There are now countless books that delve into the subject more or less objectively. We will therefore not go into detail here, but will focus more generally on nutritional behavior (Glycoplan as per Dr., Mosetter) and especially on a qualitatively and quantitatively targeted intake of micronutrients that is the subject of much controversy among experts in nutrition science.

Only a combination is useful

Based on the previously listed results one might get the impression that executives (entrepreneurs, top managers, senior staff), but also people with various disorders are dependent on a targeted intake of micronutrients, and that good nutritional behavior has much less impact on the micronutrient balance than was previously thought.

Far from it! It is true that a customized micronutrient formulation (not based on the shotgun approach and the idea that "more helps more") is critical. But our years of research show that only those who also eat a balanced diet with lots of fruit and vegetables (containing many still unresearched secondary vital substances) can also absorb the additional micronutrients.

THE POWER OF MICRONUTRIENTS

Anyone who thinks he can ease his guilty conscience with a regular intake of vitamins, minerals, and trace elements based on the motto "Eat fast food and take pills" did not comprehend the problem.

Based on our testing only 30% of the individuals we tested met the DGE's minimum daily requirement of approx. 600-800 g of fruit and vegetables (fig. 36). The individuals we tested were rarely able to meet this requirement and it completely bypasses the real working world.

We have therefore met this requirement for years now with Dr. Mosetter's Glycoplan (➡ chapter 4.7).

Fig. 36: DGE (German Nutrition Society) Requirement: 600-800 g of fresh fruit and vegetables per day

More than 70% of executives exercise and still feel mentally and physically maxed out. But they also describe the enormous difficulty of maintaining balanced nutritional behavior while frequently traveling, attending meetings and business appointments. In addition, meals are eaten at "unhealthy times", such as multi-course meals at working dinners.

Our research shows that even the many disorders result from biochemical disturbances in the functional energy metabolism. The functional energy metabolism shows considerable impairment even in "well-nourished" competitive and elite athletes. Even more serious are the effects on older people. 70% of individuals > 50 we tested take at least 3-5 medications that have a verifiably negative effect on micronutrient resorption.

Biochemistry of happiness: "Energy on prescription"

That is where quality of life can be optimized with a targeted intake of micronutrients.

As previously mentioned, the consequences of biochemical disruptions are poor sleep behavior, mental and physical performance fluctuations all the way to extreme mood changes.

The extensive and comprehensive analyses show a significant need for optimization in the area of brain-activating amino acids and other basic micronutrients in executives (entrepreneurs, top managers, executive staff) as well as in competitive/elite athletes and many of the 60,000 people we tested.

The results of our analyses in 10,270 individuals show that, for instance, the amino acids tryptophan, phenylalanine, and tyrosine have a negative effect on brain metabolism and sleep behavior. But various media outlets continue to publish data that judge vitamins and micronutrients as controversial.

For example, JAMA 2011 reported that vitamins and minerals would raise mortality in women at the average age of 61.6: multivitamins by 2.4%, vitamin B6 by 4.1%, folic acid by 5.9%, iron by 3.9%, magnesium by 3.6%, zinc by 3.0%, and copper by 18%. Only calcium showed a life-extending benefit of 3.8% (Mursu et al., 2011).

THE POWER OF MICRONUTRIENTS

In 2018, the *Deutsches Ärzteblatt*, a weekly German medical magazine, published the results of a meta-analysis in a story by Nadine Eckert titled *"Rolle der Omega-3 Fettsäuren ist unklar"* (The role of Omega-3 fatty acids is unclear). The average age of the study participants of all studies that were included in the meta-analysis was 64. They took daily dietary supplements including 220-1800 mg of eicosapentaenoic acid (EPA) and 150-1700 mg of docosahexaenoic acid (DHA). No protection from cardiac events could be verified. But the authors of the meta-analysis emphasized that additional research was needed:

> *"Our meta-analysis does not provide any evidence to support supplementation with approx. 1 g of Omega-3 fatty acids per day."*

The goal of a meta-analysis by Bolland et al. 2018, in the Lancet Diabetes & Endocrinology was to verify the significance of Vitamin D in the prevention of fractures, frequency of falls, and bone density. The meta-analysis included 81 randomized studies. Pooled analyses could not show an advantage of a high or low dose Vitamin-D supplement with respect to bone density, frequency of falls, or fracture prevention.

The authors even go so far as to say that due to the test results they see no indication for Vitamin-D compounds for locomotor-system health and demand that these results be reflected in future clinical guidelines.

In none of these studies were customized nutrient formulations analyzed, but rather fixed, patient-independent dosages. And yet a differentiated view and individualistic approach is the non plus ultra of micronutrient medicine!

4.5 Biochemical individuality: Nutrient requirement is an individual variable

Biochemical individuality is one of the most basic prerequisites for the micronutrient requirement. Therefore, when looking at and evaluating the many laboratory findings, comparison to other individuals within the age structure, possible disorders, preexisting conditions, and of course the respective nutritional behavior, is fundamental.

Nutrient requirement is influenced by many factors:

- Age, weight, height, gender;
- Pregnancy and lactation period;
- Eating habits, food intolerances;
- Medications;
- Alcohol- and tobacco use;
- Stress;
- Performance intensity: physical activity, sports;
- Utilization possibilities of available nutrients;
- State of health: illness, surgeries, chemotherapy.

National (DGE, DACH) and international standards (National Research Council, FDA) have developed reference values for nutrient intake that can prevent nutrient-specific deficiency diseases like scurvy, pellagra (Vitamin-B_3 deficiency), beriberi (various disease patterns like neuritis, edemas, etc., deficiencies in thiamine, Vitamin-B_1). These nutrient recommendations only cover the minimum requirement.

These recommendations are based on studies with healthy volunteers who don't smoke, drink alcohol, and are exposed to very little pollution. With the exception of pregnancy and lactation period, the individual nutrient requirement of each person, the biochemical individuality, is not collected in spite of added safety margins via these recommendations.

THE POWER OF MICRONUTRIENTS

Paradoxically, to date the normal nutrient requirement of older people is derived from the data on young people, regardless of the fact that in old age many factors accumulate that can lead to an inadequate nutrient supply:

- Unbalanced and inadequate food intake due to reduced basal metabolic rate and decreased metabolic activity;
- Malabsorption caused by changes in the mucosa;
- Necessary long-term medication due to chronic illnesses;
- Dental problems;
- Physical disability;
- Mental disorders: depression and dementia;
- Increased oxidative stress in old age: oxidative damage to enzymes, cell membranes, and DNA ↑; stressed cellular metabolism ↑;
- Endogenous repair processes and defense function ↓;
- Risk of free-radical diseases (KHK, cancer, cataract) ↑.

Our globally unique databank and software facilitate an optimal evaluation of blood and urine analyses, making it possible to have a differentiated view of comparable groups of persons via case history data and the results from blood and urine analyses, and create the appropriate micronutrient formulations (➡ chapter 6.2).

The results from people with comparable characteristics (age, gender, height, weight), different disorders and/or preexisting conditions, or rather the micronutrient status in a respective sport for athletes, are the evaluation criteria used for target-reference value deviations of micronutrient concentrations of the databank's many clusters. The databank holds 10,542 target-/reference values divided into 297 categories. The databank lets us identify the specific average levels of micronutrient concentrations of individual persons within the respective group.

A unique, evidence-based system has been created that makes it possible to ascertain the respective valuations of the individual person's energy status with the help of percental deviations from these target-/reference values. With the help of the thus ascertained individualized micronutrient formulations, the targeted intake and its percental changes it is now possible to make a very good assessment of where a good cellular level is achieved.

Evaluation criteria for deviations from the 10,542 target-/reference values divided into 297 categories:

- Up to -10% of target/reference values: slight deficiency.
- Up to -20% of target/reference values: distinct deficiency.
- Up to -30% of target/reference values: blatant deficiency.
- Up to +20% of target/reference values: optimization needed
- Up to +25% of target/reference values: optimal supply

4.6 Causes of increasing deficiencies

SALUTO has been able to prove in many research projects and one clinical study that even with a balanced diet the functional energy metabolism shows biochemical disturbances that are linked to activity restrictions of certain enzymes and can lead to many disorders. Currently only 30% of the population meet the DGE's (German Nutrition Society) guidelines for a micronutrient supply, whereby it is our opinion that these are far too low. Currently 82% of all tested athletes engage the help of nutritionists to balance these deficiencies. What is the reason?

The American biologist Irakli Loladze from Princeton University (2001-2003) in New Jersey already reported on initial studies in 2002, and found that increasing CO_2 emissions have a significant impact on reduced micronutrient concentrations in today's foods. He claims that the globally reported substances of content are outdated.

Increasing deficiencies in the cellular micronutrient balance could be detected in spite of a balanced diet. This matches our own test results on 100 women and 76 female students who had a very health-conscious diet and still showed micronutrient deficiencies. Loladze's studies up to 2002, describe essentially two processes:

THE POWER OF MICRONUTRIENTS

1. The increased CO_2 emissions promote plant growth and thus increase crop yields, but that happens at the expense of valuable substances of content (trace elements, other micronutrients).
2. The higher CO_2 values limit water evaporation from plants. Since less water evaporates from the leaves the ground also absorbs less water. This reduced water balance leads to decreased availability of iron, magnesium, and zinc from the ground.

Nowadays we must assume a significant reduction of substance of content in our food. If an experiment measured losses on such a scale in only six months, the actual losses over the past decades must be far greater. Today's vegetation has 30% more exposure to carbon dioxide than it did 200 years ago (Zhu et al., 2018), and to date no end to the increase in CO_2 emissions is in sight.

This seems to be one of the main reasons that physically active people in particular cannot achieve an adequate micronutrient supply even with a balanced diet. Next to the severe consequences of micronutrient losses listed below due to the greenhouse effect, there are additional disturbances such as improper preparation and storage-related losses. All of it is a problem of industrialized food that is further aggravated by a definite increase in food intolerances.

The conclusion immediately suggests itself that nowadays we must go on the assumption that there is a reduction of substance of content in our food, which different scientists however continue to deny in their "lobbying work".

In a pilot project at the TU Braunschweig scientists simulated the greenhouse effect in the year 2050 (fig. 37). The results show a significant reduction of micronutrients (zinc, selenium, etc., as well as protein concentration) in the studied plants, with the exception of calcium. Plants will grow more quickly but will contain significantly fewer micronutrients.

A description of the FACE experiments can be found at https://www.thuenen.de/de/thema/klima-und-luft/experimentelle-klimawirkungsforschung/wie-das-face-experiment-funktioniert/ (Johann Heinrich von Thünen-Institut, Braunschweig/Germany).

Fig. 37: The CO2 increase changes plant chemistry.

The previously described effects have been confirmed by further studies. In 2018, Smith and Myers researched the nutrient content of 225 plants in 151 countries. Currently the CO_2 concentration lies at 440 ppm. An increase to 550 ppm by 2050 is projected. This means that by 2050, wheat, rice, and corn will contain 17% less iron, zinc, and protein than they do today. This is particularly problematic for Southeast Asian countries where rice is the primary nutrient source for more than 2 billion people (https://advances.sciencemag.org/content/4/5/eaaq1012, Zhu et al., 2018).

In a study titled: *This century's carbon dioxide content will change the protein, micronutrient and vitamin content of rice grains and will potentially have health implications for the poorest rice-dependent countries*, scientists from China, Japan, Australia, and the US were able to show via an in-situ FACE (free air CO2 enrichment) experiments (fig. 37) that the protein, iron, and zinc content as well as the amount of

B1, B2, B5, and B9 vitamins significantly dropped during a CO2 increase, while the Vitamin-E content rose.

4.7 The modified Glycoplan as per Dr. Mosetter

The Glycoplan is a nutrition concept that was successfully developed in recent years by Dr. Kurt Mosetter (Mosetter et al., 2018). Many years of personally experiencing many different cultures have made this concept so successful. Dr. Kurt Mosetter is a physician and founder of Myoreflex therapy. His therapeutic focus areas include pain and neuromuscular therapy as well as metabolic and nutritional medicine. He also is the director of "interdisciplinary therapies" in the cities of Constance, Gutach, and Herrenberg.

The purpose of the Glycoplan is to provide a healthy diet, to supply the human body with adequate nutrients, achieve improved wellness, and stay healthy long-term. The main focus, on the one hand, is reducing bad carbohydrates, fats, allergens, and other harmful substances, and on the other hand, the intake of important nutrients such as vitamins, minerals, amino acids, and secondary plant substances. The typical diet thus often consists of:

- Sugar and white flour products;
- Trans-unsaturated fatty acids;
- Artificial flavors, preservatives and sweeteners, as well as additives and is burdened with
- Antibiotics and pesticides.

The targeted intake of these individual substances is very important to many processes of the human body. The overall goal is to achieve a balanced diet between the three carriers of energy protein, fat, and carbohydrates. Furthermore, when choosing carbohydrates, we differentiate between so-called good and bad carbs. The Glycoplan is basically universally applicable, whereby it can be adjusted for people with obesity, allergies, celiac disease, gluten intolerance, lactose intolerance, ADHD, and competitive athletes.

An easy-to-understand traffic light system and adjusted timing for food intake help to make the dietary change. The Glycoplan is available online at http://www.myoreflex.de/uebersichtsseite-ernaehrung/glycoplan last access: July 12, 2020).

4.7.1 The Glycoplan traffic light system

Glycoplan foods are divided into the following three categories or rather three traffic light colors:

- Green means recommended;
- Yellow includes foods that "should only be eaten occasionally", and
- Red is the category of foods that should largely be avoided.

Green foods

The group of green food includes largely vegetarian and vegan foods like vegetables, lentils, legumes, nuts, berries, herbs, and various mushrooms. These foods are supplemented with things like fish, particularly wild salmon, halibut, herring or tuna, as well as oat, almond, and coconut milk products. Furthermore, the green list includes moderate amounts of organic goat or sheep milk products such as cheese or yoghurt as well as venison and lamb. Beyond the color there is additional information about the individual foods.

For instance, fruit should only be eaten before 2 pm, and generally not more than 50-100 g, depending on the fruit.

Galactose, a simple sugar, as well as glucose, a carrier of energy, are very important during the dietary change. Galactose is found in lentils, berries, and chickpeas, but especially in milk and milk products. The diet allows an intake of up to 25 g/day.

The advantage of galactose is its insulin-dependent transport into the cell. This means that a glucose deficiency, for instance in cases of beginning insulin resistance, can be balanced with a galactose intake, thereby continuing to provide the cells in brain, liver, and muscles with sufficient energy.

Galactose, which is available as a distillate in powder form, is therefore very helpful during the dietary change and the accompanying sugar withdrawal (Mosetter et al., 2018).

Yellow foods

Next to some fruits like pineapples, apples, mangoes, melons, peaches, and quinces, which should only be eaten in moderation and before 2 pm, the group of yellow foods also includes some dried fruits with an indication of quantity that should not be exceeded. For example, the daily intake of dried cranberries should not exceed 10 g/day.

Organic meats and skinless poultry are also included in the "yellow" food group. Flour- and grain products like potatoes, red and black rice, oats, spelt, einkorn wheat, and emmer flour are also on the "yellow list". Beverages such as fruit teas or fresh fruit juices and the occasional dry, high-quality wine are also on this list. When it comes to foods that contain carbohydrates, the rule is that they should be eaten before 2 pm (Mosetter et al., 2018)

Red foods

The group of red foods incudes foods that should ideally be avoided. That means they are sugary, industrially processed or difficult-to-digest foods. On the list are fruits like plums, grapes, and prunes, which are high in fructose, as well as so-called fast food like pizza, burgers, kebabs, chips, etc.

Moreover, soft drinks, sweets, sugar, alternative sweeteners, and wheat products like pasta and white bread should be avoided. Sweet potatoes should be avoided due to their "high glycemic index". Due to their burden of pollutants, hormones, and antibiotics pork, shrimp, as well as cheap turkey meat should also largely be crossed off the menu.

4.7.2 Summary of nutritional aspects

We have modified the Glycoplan and have been able to use it very successfully in various studies and research projects. This Glycoplan combined with a customized micronutrient intake has verifiably resulted in the following changes:

- Improved metabolic performance,
- Increased physical and mental performance capacity,
- Activation of the microbiome and thus the immune system,
- Protection from diseases (of civilization).

What many diseases of civilization have in common is an imbalance of the glucose metabolism processes with consecutive insulin resistance. As a result, uric acid as well as liver values and blood pressure go up. Sugar and free fatty acids trigger inflammatory responses and a "leaky gut" (➡ chapter 3.2). These inflammatory responses initially smolder undetected.

In the long-term, they lead to inflammatory diseases such as metabolic syndrome (obesity, diabetes, type 2 diabetes, high blood pressure, lipid metabolism disorder), arteriosclerosis, heart attack, stroke, dementia, multiple sclerosis, or Parkinson's disease. What all of these diseases have in common is that they are often initiated by initially "silent", harmless inflammatory reactions that are primarily reversible.

A successful nutritional strategy leads to improved metabolic performance, increases physical and mental performance capacity, supports the immune system, and protects from diseases of civilization.

THE POWER OF MICRONUTRIENTS

In a functioning metabolism, the insulin level drops at night. Fat oxidation, hormone production (serotonin, melatonin), growth and repair processes can proceed. That is why the timing of food intake is critical: When sugar is consumed in the evening the pancreas releases insulin. As a result, nightly fat oxidation does not function and hormone synthesis of growth and regeneration hormones is inhibited. These factors then result in poor sleep.

At this point it is important that the approach take a holistic view of the human body. It can be compared to an orchestra, many individual sub-systems playing together. This must be looked at and optimized both at the

- **Microlevel:** Minerals and vitamins, all organs, and the entire metabolism as well as the
- **Macrolevel:** Lifestyle including individual components such as sleep, stress, exercise, social life.

PLEASE NOTE

Lifestyle changes, a healthy nutritional strategy (Glycoplan), and a targeted, customized dose of micronutrient supplement together should increase wellbeing.

Superfoods

Nowadays everyone talks about *superfoods*. However, we, the author and his team, do not just think of exotic fruits as such, but superfoods, for different reasons, are:

- Berries,
- Fresh spices,
- Legumes,
- Germ buds and sprouts,
- Coconut,
- Omega-3 fatty acids,
- Secondary plant substances,
- Wild herbs.

Many vegetables, low-fructose fruits, and fresh herbs are untreated and plant-based. Valuable fats and proteins can be found in nuts, kernels and seeds, in avocados, legumes, fish, organic eggs, and oat milk. Valuable carbohydrates come from preferably gluten-free whole-grain products such as quinoa and ancient grains like millet and buckwheat.

5 THE ROLE OF INDIVIDUAL MICRONUTRIENTS: HOW MICRONUTRIENTS REALLY HELP

5 THE ROLE OF INDIVIDUAL MICRONUTRIENTS: HOW MICRONUTRIENTS REALLY HELP

Here we will choose just a few examples and skip the detailed biochemical explanations. An overview of the individual micronutrients with respect to dosage and their effect can be found in tab. 37, 38 and 39 in the appendix.

5.1 Amino acids: Life's building blocks

Proteins are basic building blocks of all of the body's cells and control all biochemical processes in the body. As the basic building block of muscle fiber and as the structural protein of cartilage, bones, tendons, and skin, they are the body's most important structural elements. The organism undergoes continuous synthesis and degradation of proteins. Amino acids are the basic building blocks of proteins.

Our amino acid storage is small, only about 120 g, and is located in blood plasma and primarily in the muscles and in cellular structures. A daily, high-quality protein supply from food is indicated because the organism cannot store sufficient quantities of amino acids (tab. 2).

The structural and functional range of amino acids illustrates their great importance to various somatic and mental processes of the human organism. At this point we will skip the many functions and tasks of the individual amino acids. This applies particularly to the important amino acids phenylalanine, tryptophan, tyrosine, which are especially important for brain metabolism and the emotions of athletes.

Amino acids, like arginine, proline, methionine, and glycine that are especially important for the elasticity and maintaining of function of the myofascial system, support the fibroblasts during collagen-fiber synthesis in the extracellular matrix. The extracellular matrix consists of fibers such as collagen fibers, reticular and elastic fibers, and the base substance consisting of glycosaminoglycan, proteoglycan, glycoprotein, and water.

The fasciae consist of fibroblasts and the surrounding matrix. Fibroblasts are connective-tissue cells. They produce the collagen fibers, that make up most of the matrix. One could say: They build their home with collagen (fig. 38).

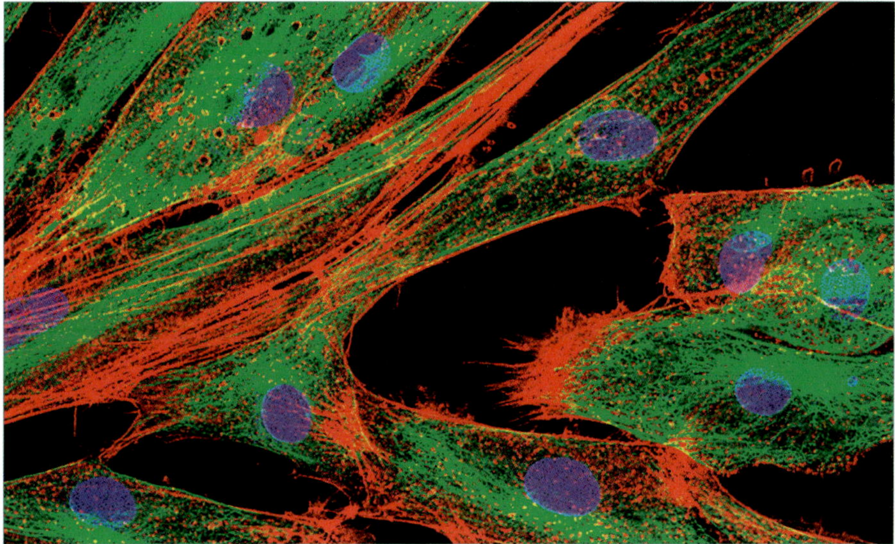

Fig. 38: Fibroblasts

Mental and physical performance capacity and the myofascial system are greatly affected by the connective-tissue structures and the many demands placed on them. A good supply of high-quality collagen peptides guarantees a well-functioning brain metabolism, tendon, muscle, cartilage, and bone structures, the complex immune system, and the elasticity of the entire myofascial system. An optimal supply of amino acids thus guarantees performance and protection from injury.

5.1.1 Essential and nonessential amino acids

The human body is incapable of producing eight out of the 20 amino acids. Essential amino acids must be ingested from food. Meanwhile some semi-essential amino acids have been identified whose regular intake is necessary to take on important metabolic functions.

To what extent their typical separation into essential and non-essential amino acids will be retained is in question since some research papers have shown that the boundary between the two groups can be fluid. We were able to show that even non-essential amino acids can be essential in certain situations. According to the current definition, eight of the 20 amino acids are indispensable (essential):

- **Essential amino acids** include: isoleucine, leucine, lysine, methionine, phenylalanine, threonine, tryptophan, valine.
- Depending on the stage of life (newborn, growth, convalescence), or because they synthesize from essential amino acids, **semi-essential amino acids** include: arginine, cysteine, histidine, tyrosine,
- **Non-essential but still important amino acids** include: alanine, asparagine, aspartic acid, glutamine, glutamic acid, glycine, proline, serin.

5.1.2 The tasks and functions of some amino acids

Amino acids in natural proteins are always L-amino acids. The L refers to the molecule's spatial structure. With few exceptions, the body can utilize only L-amino acids. Whenever we refer to amino acids here, we are always talking about the L-amino acids (e.g. glutamine = L-glutamine).

High-quality amino acids complete important tasks in the organism, including during

- Energy metabolism: Creatine phosphate synthesis; buildup of glycogen stores;
- Facilitates burning of fatty acids;
- Improved mental and muscle regeneration;
- Optimal protection from injuries and immune stabilization;
- Buildup of bradytroph tissue structures.

PLEASE NOTE

Amino acids guarantee mental and cognitive performance capacity.

Mental and physical performance capacity is no coincidence. A good mood, a great performance, and creativity all depend on an adequate amino acid supply that affects the functional energy metabolism. Mental and physical stress can significantly increase the demand.

As nutrients, amino acids have a direct impact on your personal wellbeing. With the right food choices, you can influence your own status and achieve "a good mood".

Euphoria and optimism are mirror images of your amino acids. An optimal supply with the aforementioned elemental amino acids will let misery and a bad mood quickly fade away.

Phenylalanine

This is the precursor of catecholamine, dopamine, adrenaline, and noradrenaline. Disruptions to the catecholamine metabolism are involved in increasing exhaustion. L-phenylalanine has a mood and attention-affecting property. All tested subjects who complained about a state of exhaustion showed definite deficiencies in this amino acid.

Tryptophan: The amino acid that lets us sleep

Tryptophan is the precursor to the neurotransmitters serotonin and the hormone melatonin. A serotonin deficiency is being debated as a significant factor in sleep disorders and depressive mood disorders. A serotonin deficiency can be balanced with an additional intake of tryptophan.

In contrast to serotonin, tryptophan can overcome the blood-cerebral barrier and reach the brain. There it is converted to serotonin.

It is extremely difficult to cover the tryptophan content with a balanced diet. Tryptophan deficiencies slowly increase over time and the individual feels increasingly exhausted, sleeps badly, and has mood swings. Tryptophan should never be taken arbitrarily at night before going to sleep. This should only be done after a proper diagnosis. A dose of tryptophan can be counterproductive in people taking certain medications, such as antidepressants, or who suffer from chronic diseases, because it is biochemically metabolized into toxic substances. In cases of inflammatory responses 5-HTP is given, which is an amino acid that, when bypassing the blood-cerebral barrier, has no negative effects.

We have regularly and successfully used tryptophan for sleep disorders and increasing mood swings. An increase in the serotonin as well as the melatonin level in the brain is apparently needed for a therapeutic effect. Dosages of 100-400 mg/5-HTP are recommend for sleep disorders, depending on blood test results.

PLEASE NOTE

Tryptophan's resorption is slow. It should be take approx. 2 hours before going to sleep. The effect only becomes noticeable after about 4-5 days with increasing therapy duration. It is also suitable for chronic insomnia and can be used for physical withdrawal in cases of sleeping-pill dependency.

When diagnosing the status of the amino acid tryptophan, additional blood tests are necessary in some cases to achieve better informational value for the individual.

Tab. 2: General nutritional significance of individual amino acids

Amino acids (essential, semi-essential, non-essential)	Special properties	Particularly high content in:
ISOLEUCINE (essential)	• Promotes protein synthesis. Emergency energy carrier for muscle cells. • Counteracts excessive serotonin production.	• Lactalbumin • Casein • Meat protein • Egg protein • Hazelnut protein
LEUCINE (essential)	• Promotes protein synthesis in muscles and liver (protein anabolism). • Inhibits degradation of muscle protein (catabolism). • Emergency energy carrier for muscle cells. • Counteracts excessive serotonin production (less fatigue during high-stress situations). **Caution:** As a nutritional supplement leucine should only be taken with isoleucine and valine, otherwise the protein synthesis can be impaired.	• Whey protein • Oat protein • Corn protein • Millet protein • Egg protein • Casein free in cocoa • Hazelnut protein
LYSINE (essential)	• The most value-limiting amino acid in vegetarian foods. • Forms indigestible products when heated to high temperatures. Hence baked goods are usually low in lysine. • Source material for endogenous carnitine. • Intensive fertilization lowers lysine content and thus the value of wheat protein! • The extremely low lysine content of cornflakes makes them practically worthless from a nutritional standpoint. • Increases the effects of arginine.	• Lactalbumin, casein • Egg protein • Meat protein • Soy protein • Potato protein • Amaranth protein • Wheat protein • Lentil protein free in potato water
METHIONINE (essential)	• Provides sulphur for many syntheses of endogenous substances. • Provides methyl groups for syntheses. • Particularly high demand during anabolic phases. • Improves wound healing. • Facilitates production of cysteine and taurine in the body. • The sum of methionine and cysteine is critical in the assessment of proteins. • "Sulfuric acid" is created during methionine degradation; more alkaline foods should be consumed (vegetables, fruit, magnesium citrate) to prevent acidosis from food that his high in methionine.	Applies to the sum of methionine and cysteine: • Egg white protein • Whole egg protein • Fish protein • Liver protein • Oat protein • Brazil nut protein • Whole corn protein

Amino acids (essential, semi-essential, non-essential)	Special properties	Particularly high content in:
PHENYLALANINE (essential)	• Precursor substance to the activating neurotransmitter dopamine • A targeted dose can reduce production of the fatiguing neurotransmitter serotonin. • Since dopamine is produced via the intermediate stage tyrosine, the sum of both amino acids is needed for the assessment of a protein.	Applies to phenylalanine and tyrosine: • Casein • Hazelnut protein • Whole rice protein • Peanut protein • Egg white protein
THREONINE (essential)	• Is easily used for energy production during severe physical strain. • Threonine demand is particularly high during anabolic phases.	• Whey protein • Egg yolk protein • Pea protein • Wheatgerm protein • Beef
TRYPTOPHAN (essential)	• Precursor of the calming neurotransmitter serotonin • A targeted intake promotes sleep-readiness. • Important for vitamin synthesis in the body (niacin) • Tryptophan is destroyed during the technical protein hydrolysis and is thus missing in gelatin hydrolysates.	• Lactalbumin (as part of whey protein) • Cashew protein • Whey protein • Egg white protein
VALINE (essential)	• Emergency energy carrier for muscle cells • Counteracts excessive serotonin production	• Lactalbumin, casein • Oat protein • Meat protein • Egg protein • Hazelnut protein • Whole rice protein

Amino acids (essential, semi-essential, non-essential)	Special properties	Particularly high content in:
ARGININE (semi-essential)	• Important for liver metabolism (urea production, ammonia decomposition) • Particularly important during the early regeneration phase after intense physical strain. • Improves anabolic utilization of milk protein (low in arginine) for 100 g casein at least 4 g, for 100 g whey protein at least 5 g, and for 1 l milk 1.5 g arginine. • In the body arginine is quickly converted to ornithine and vice versa. That is why arginine can be largely substituted with ornithine. From a nutrition-physiological point of view the sum of arginine and ornithine is critical. • In higher quantities improved protein synthesis (anabolism) via increased release of growth hormones. When taken on an empty stomach, even just 3 g can be effective. • Promotes decreasing hair growth, particular with local administration. • Supports the immune system, e.g. in overtrained athletes – here activation of "dormant herpes" is possible and in some circumstances bothersome. • Improves fat metabolism and can lower cholesterol concentration in the blood. • Higher quantities can have a dehydrating effect (then divide intake into smaller portions). • Precursor for nitric oxide (NO), which has an antioxidant function and promotes circulation by widening peripheral vessels.	• Almonds 14.8 g • Walnuts 14.6 g • Peanuts 13.8 g • Coconut 12.6 g • Meat-/fish- and soy protein 7-8 g/100 g • Whole rice protein 8.3 g/100 g • Oat protein 7.3 g/100 g

Amino acids (essential, semi-essential, non-essential)	Special properties	Particularly high content in:
CYSTEINE (semi-essential)	• Important for the secondary structure of many proteins due to the forming of sulfur bridges • An extremely conditional high cysteine content is the cause for curly or frizzy hair. • Supplementation with cysteine is said to promote hair growth. • From a nutrition-physiological point of view it depends on the sum of methionine and cysteine as up to 2/3 of the methionine demand can be met in the form of cysteine (also see methionine!). • In the blood cysteine occurs primarily in the form of the double molecule cysteine. • Cysteine is not very water-soluble and thus marginally usable in fluid preparation. • Cysteine is necessary for the synthesis of glutathione as it is important for the protection of cells against free radicals (e.g. during athletic activity).	• Egg white protein • Oat protein • Whole corn protein
HISTIDINE (semi-essential)	• Important for the production of the blood pigment heme. • Of all amino acids it is the easiest to be discharged in urine. Since zinc bound to histidine is also lost here, a high histidine intake should be avoided. • Is said to promote blood-clotting as a supplement. • Unlike other amino acids only about 60% is resorbed in the colon. • Can promote cell protection.	• Banana protein (8%) • Tuna protein • Mackerel protein • Beef protein
TYROSINE (semi-essential)	• Precursor of the activity-increasing neurotransmitter dopamine. • Targeted intake can have a stimulating effect. • From a nutrition-physiological point of view it depends on the sum of phenylalanine and tyrosine as tyrosine is produced from phenylalanine. • In people with kidney disease the body's own production of tyrosine can get so low that tyrosine supplementation is important. • Tyrosine has very poor water-solubility and is therefore marginally usable in fluid preparations.	• Casein • Whole milk protein • Pea protein • Egg yolk protein • Peanut protein • Bean protein

Amino acids (essential, semi-essential, non-essential)	Special properties	Particularly high content in:
ALANINE (non-essential)	• In catabolic situations it serves to remove amino groups from muscles and liver (urea synthesis). • In cases of glucose deficiency or during stress metabolism it serves in glucose production in the liver (gluconeogenesis). • Significant amounts of alanine are used after hour 2 of intensive endurance loads. • An alanine replacement is recommended after endurance loads longer than 1 hour. • Alanine can also promote protein synthesis in the liver.	• Gelatin (9.8%) • Whole corn protein • Beef protein • Egg white protein • Pork protein • Rice protein • Soy protein • Oat protein
ASPARAGINE (non-essential)	• Has a close metabolic relationship with aspartic acid. • In the human organism it is apparently of relatively low significance, unlike aspartic acid. • As a free amino acid in many fruit, berry, and vegetable juices; apple juice can contain approx. 1 g/l.	
ASPARTIC ACID (non-essential)	• Has a key function in energy metabolism, the liver etc. • Supports ammonia detoxification via the urea cycle in the liver. • An intake during the early regeneration phase is particularly important to normalize ammonia concentration. • Is produced in the body via asparagine breakdown, etc. • Is free in many fruit juices and vegetables (together with asparagine): passionfruit 1.6 g/l. • Food charts list aspartic acid and asparagine.	• As aspartic acid and asparagine • Potato protein (20.7%) • Coconut protein (17.1%) • Alfalfa (12.3%) • Peanut protein • Egg white protein • Meat protein

Amino acids (essential, semi-essential, non-essential)	Special properties	Particularly high content in:
GLUTAMINE (non-essential)	• In terms of quantity, it is the most important free amino acid in the body. • Is produced from glutamic acid through ammonia incorporation. • Most important energy carrier for some cells, such as, for example, mucosal cells (in the small intestine) and for immune-system cells). • Stimulates glycogen synthesis (as an ingredient in wheat flour it is important for use of "carb-loading" before a competition). • Improves protein synthesis in catabolic metabolism situations. • Regulates muscle protein synthesis (likely by stabilizing cellular water balance). • Essential in catabolic metabolism situations. • Some important foods also contain a lot of glutamine (more than glutamic acid). The glutamic acid content in milk protein of 20-24% listed on food labels is actually 2/3 based on glutamine that only became glutamic acid during the analysis process. • Some foods also contain significant amounts of free glutamine: tomatoes (0.65%), potato water and other vegetables. • Large amounts of glutamine can be lost during intensive athletic activity. Quick replacement helps regeneration and secures the anabolic training effect. • Glutamine is necessary for glutathione production, which is important for the protection of cells against free radicals (e.g. athletic activity).	• Wheat protein (up to 30%) • Oat protein • Casein • Whey protein

Amino acids (essential, semi-essential, non-essential)	Special properties	Particularly high content in:
GLUTAMIC ACID (non-essential)	• Important reloading point for amino nitrogen in the metabolism. • Is produced during degradation of some amino acids as an intermediate stage, particularly with proline, histidine, arginine, and ornithine. • During intense endurance loads a large part of the free glutamic acid store is lost in spite of regeneration. Therefore, an intake of this amino acid is sensible after endurance loads. • Can bind ammonia during glutamine production and transport it to the liver, which turns it into urea and glucose, supports arginine in the ammonia-lowering effect. • Promotes protein synthesis – likely the reason for the high glutamic-acid content in milk. • Most used seasoning in the world in the form of sodium salt (glutamate). • The previously-assumed increased brain performance is very controversial. • Dr. Nöcker described a positive effect on physical performance as early as 1953. • Excessive individual doses can cause nausea in very sensitive individuals, likely due to a Vitamin B6 deficiency (Chinese Restaurant syndrome). • For reasons of analytical technique only the sum of glutamic acid and glutamine is known in most foods. • Some foods contain primarily pyroglutamic acid (potato), which has a lower water content.	• White flour protein • Whole wheat protein • Casein • Potato protein • Hazelnut protein • Pork protein • Whole rye protein • Whey protein • Beef protein • Soy protein

Amino acids (essential, semi-essential, non-essential)	Special properties	Particularly high content in:
GLYCINE (non-essential)	• As the most basic of all amino acids, it is a universal supplier of amino groups for the synthesis of other substances, e.g. blood pigment hemoglobin. • Important for the synthesis of connective-tissue protein. • High glycine demand during anabolic phases. • Danger of degradation of connective-tissue protein with an insufficient intake. • High glycine intake impedes the protein-degrading enzyme cathepsin D and decreases the degradation of connective tissue during catabolic situations (preserves the barrier function of connective tissue against invading germs) • Glycine has cell-protecting properties when there is a temporary shortage of oxygen.	• Gelatin (23.8%) • Beef protein • Liver protein • Peanut protein • Oat protein
PROLINE (non-essential)	• Proline is particularly important for the synthesis of connective-tissue protein: Hydroxyproline released during connective-tissue degradation cannot be reused for synthesis and must be excreted with the urine. • During an energy shortage, for instance when fasting for an extended period of time or during athletic endurance performances, large amounts of proline are used for energy production. • Fruit juices contain substantial amounts of free proline, e.g. up to 2.5 g per liter in orange juice.	• Casein • Whole-milk protein • Wheat germ protein
SERINE (non-essential except with kidney failure)	• Normally it can be easily produced from glycine (primarily in the kidneys) while using methyl, e.g. from methionine, and vice versa. • Due to the close link between serine and glycine, the sum of both amino acids is important from a nutrition-physiological point of view.	• Egg yolk protein • Egg white protein • Casein • Whey protein • Oat protein • Corn protein

Amino acids (essential, semi-essential, non-essential)	Special properties	Particularly high content in:
TAURINE (amino-acid-like)	• Stable end product of sulfurous amino-acids metabolism. • Important for the stabilization of cellular water balance and thus protein synthesis. • Vital to the development of infant brains. • Has cell-membrane-protecting properties and is thus useful during major athletic loads. • Acts as a free radical scavenger and protects cells from damage during intensive loads. • Has a regulating function in the heart muscle and in Japan is thus thought to protect the heart from stress (is consumed there in large quantities as heart- protection lemonade). • Promotes production and effectiveness of bile (taurocholic acid) as an emulsifier during fat digestion. • Prevents overstimulation of nerves from caffeine, thereby increases the safe consumption of caffeinated beverages (also in sports). • Can have a stimulating effect! • Can increase transportability of fat-soluble foreing substances in the body and promotes their excretion.	• Meat extract • Meat • Mussels

5.1.3 The effects of amino acids on pain symptoms and general wellness

Preliminary intervention studies show verifiably positive effects on the myofascial system with a targeted intake of amino acids. With respect to preventative and therapeutic aspects related to amino acids, research shows great potential: For example, a master project about the importance of amino acids with respect to pain symptoms was completed as part of a master thesis on micronutrient therapy and regulatory medicine at FMH Bielefeld.

The goal of the master project was the empirical study of the impact of amino acids on pain symptoms and general wellbeing via an experimental double-blind study. The intervention took place over an eight-week period. During this period, test subjects were instructed to ingest 16 pellets a day of a hydrolyzed pea protein of the amino acid complex or the placebo. The preparation used for the control group was microcrystalline cellulose in pill form. The intake was divided into three individual doses spread out over the day (dosage of amino acids, tab. 3). Test subjects were asked to take 2 x 6 pellets in the morning and afternoon as well as 1 x 4 pellets in the evening, each dose prior to a main meal.

Tab 3: Comparison of dosage of the amino-acid study preparation and the corresponding share of the daily requirement as per DGE (German Nutrition Society)

Amino acid	Amount per pellet	Amount per per daily dose (16 pellets)	Percentage of daily requirement of a person weighing 70 kg
L-Leu	92 mg	1.472 mg	54,5 %
L-Val	82 mg	1.312 mg	72,9 %
L-Ile	69 mg	1.104 mg	78,9 %
L-Lys	69 mg	1.104 mg	52,6 %
L-Phe	63 mg	1.008 mg	96,0 %
L-Thr	56 mg	896 mg	81,5 %
L-Met	35 mg	560 mg	80,0 %
L-Arg	22 mg	352 mg	–
L-Trp	15 mg	240 mg	21,7 %
$\sum$	**503 mg**	**8.048 mg**	**62,5 %**

All test subjects trained 3-4 hours a week for many years under medical and sport-scientific supervision. A combination of functional training, strength, and endurance training on apparatus with personalized target heartrate. All test subjects had problems in the active and passive locomotor system for years, which improved over the years with exercise.

The results of the master thesis contradict previously postulated statements from experts that the targeted intake of amino acids should be dismissed and cannot lead to a verifiably improved quality of life. To test the structural equality of the test groups the collective test subjects were tested with respect to their key tendency. Compared were arithmetic group means of the baseline data compiled during the preparticipation screening.

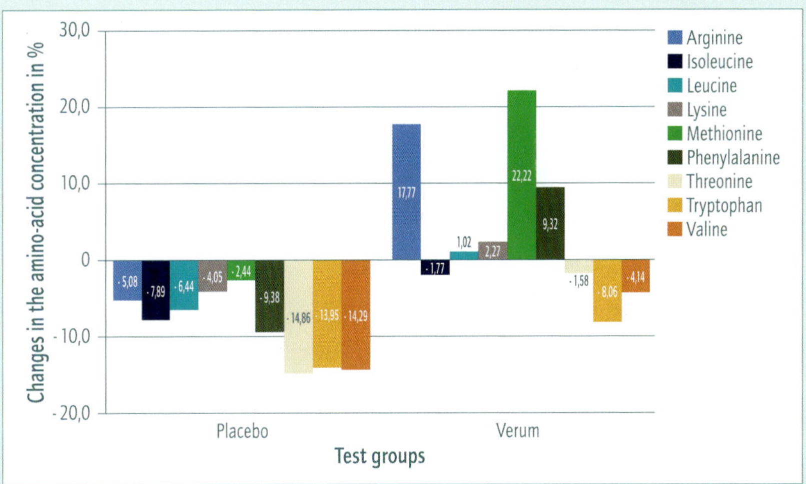

Fig. 39: Percentage change to the amino acid concentration of T1 and T2 in the placebo and verum groups

In addition, the parameter nutritional behavior was also tested within the groups for significant differences from T1 to T2. Fig. 39 illustrates and describes relevant descriptive statistics for both groups of test subjects as well as the evaluation results from the statistical tests (Herbst 2020).

It became apparent after the eight-week intervention period that test subjects in the verum group experienced a significant positive effect with respect to pain sensation as compared to the placebo group ($p < 0.05$), which does not mean that the observations cannot be considered relevant.

During the master project, the dualism of clinical and statistical relevance is already made clear from the start. The subjectively compiled data is supported by the objective variables of the amino acid measured values: The intake of the amino acid preparation was measurably reflected in the changes to the amino acid levels of the verum group as compared to the placebo group, particularly for the amino acids methionine, phenylalanine, tryptophan, and valine ($p < 0.05$; fig. 39).

In summary this work highlights the finding that amino acids can have a positive effect on pain symptoms. Moreover, it shows trends within the scope of the intervention that point to an improved overall sense of wellbeing. There is a need for future research to validate the obtained results. The sample size in particular should be larger in future studies and the intervention period should be longer. Additionally, invitro studies are necessary that focus on the biochemical modes of action on which the obtained results are based and address the issue of causality. Further details of the master project can be obtained from the FHM publication series, Bielefeld Heft 12, Meilensteine in der Gesundheitsmedizin (https://www.fh-mittelstand.de/fileadmin/pdf/Schriftenreihe/Schriftenreihe_12_web.pdf; Last access on July 25, 2020).

5.2 The importance of Omega-3 fatty acids

The HS-Omega-3 index is the benchmark for a good mood as it measures the activity of the serotonin metabolism.

Today the biochemical significance of Omega-3 fatty acids for the preservation of basic functional sequences in our metabolism is controversial. Omega-3 fatty acids are part of the cell membranes and thus are responsible for membrane permeability. Fig 40 taken from Prof. Dr. von Schacky's lecture at the FHM Bielefeld illustrates the complexity of the cell membrane's many functions which can verifiably be improved with a good supply of Omega-3 fatty acids.

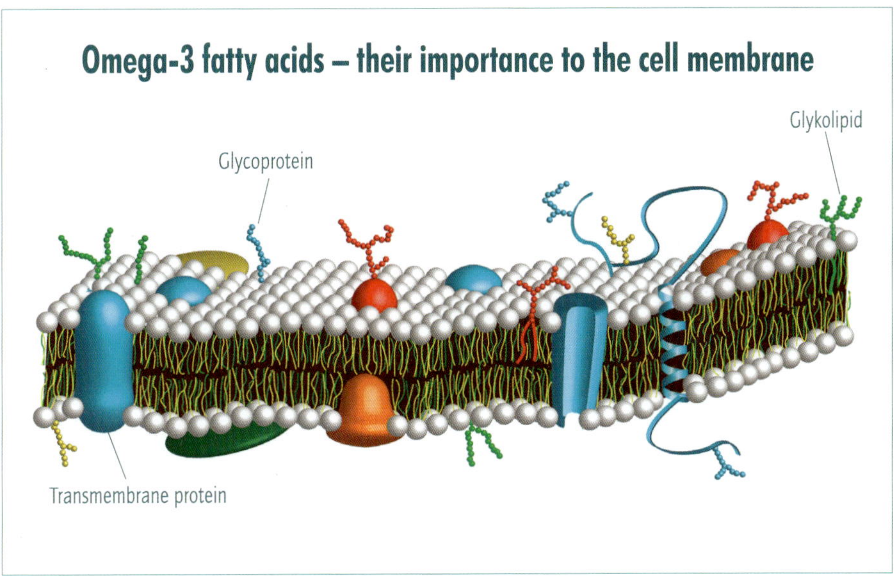

Fig. 40: Omega-3 fatty acids – their importance to the cell membrane

Omega-3 fatty acids are precursors to prostaglandins and leukotrienes with an anti-inflammatory, anti-thrombotic, vasodilatory, anti-arrhythmic effect, and also lower LDL- and elevated HDL-levels. Omega-3 fatty acids therefore have a cardioprotective effect (vascular inflammation decreases, plaque buildup diminishes), immune competence increases causing tumor risk to decrease, wound repair improves as do vision or sensory disturbances. Omega-3 fatty acids also impact brain and neural development, serotonin-/dopamine receptor synthesis, as well as retina development.

The most important Omega-3 fatty acids include α-linolenic acid, eicosapentaenoic acid (EPA), and docosahexaenoic acid (DHA). α-linolenic acid is the "starting fatty acid" in the Omega-3 fatty-acid metabolism and is found in ferns, mosses, and some plant-based oils such as linseed and rapeseed oil. The human organism can barely (about 5-10%) convert the shorter-chain α-linolenic acid into EPA and DHA. For this reason, the health enhancing effects of EPA and DHA cannot be achieved with the intake of plant-based α-linolenic acid.

The following foods contain Omega-3 fatty acids: fatty cold-water fish (salmon, mackerel, herring, sardines, tuna), microalgae, linseed oil, rapeseed oil, nut oil, sesame oil, and soybean oil.

EPA and DHA are essential elements of every cell membrane (fig. 40) and therefore are significantly responsible for your health. They strengthen:

- the immune system;
- have an anti-inflammatory effect;
- support the supply of oxygen to the organs;
- lower elevated blood lipids;
- increase concentration and mental performance capacity.

Meanwhile there is a so-called *HS-Omega-3 index* that is an indicator for serotonin production in brain metabolism, etc. (information about the HS-Omega-3 index: https://www.omegametrix.eu/index_patienten.php). When the percentage of these two Omega-3 fatty acids (EPA, DHA) in erythrocytes is too low and the Omega-3 index is < 8%, it is considerably more likely for increased mood swings to occur.

In competitive and elite athletes an HS-Omega-3 index of > 12% can verifiably have an anti-inflammatory effect after intense training and competition phases that provides good protection from muscle injuries. In cases of increasing states of exhaustion, a therapeutic effect can be achieved with concentrations of at least 840 mg EPA and 560 mg DHA.

Therapeutic effects can be seen in the evidence-based, retrospective study with 160 children with ADHD (➡ chapter 6.3.1). But an HS-Omega-3 index of > 14.5% is also recommended for other illnesses like, for instance, rheumatoid arthritis. Many years of research have shown this to be the case.

A current master project on the subject of micronutrient therapy and regulatory medicine studies the relationship between the HS-Omega-3 index and the absorption of micronutrients within the scope of a retrospective intervention study. Its primary focus is the possible improvement of cell-membrane elasticity with a good supply of Omega-3 fatty acids. The project's objective was to verify that the targeted intake of Omega-3 fatty acids (measured by the HS-Omega-3 index) and the absorption of micronutrients, using the example of ferritin, zinc, selenium, arginine, folic acid, can also improve the absorption of micronutrients and consequently the supply status.

A total of 99 test subjects, whose anthropometric data is listed in tab. 4, were chosen from the SALUTO database and divided into three groups.

Tab. 4: Anthropometric data of 99 test subjects as part of a retrospective, evidence-based intervention study

Group	N	Age in years (∅, sd)	Height in m (∅, sd)	Weight in kg (∅, sd)	BMI (∅, sd)
A	34	54,2 ± 5,8	1,73 ± 0,1	80,41 ± 11,3	26,75 ± 2,4
B	29	50,5 ± 3,4	1,74 ± 0,1	80,59 ± 12,7	26,40 ± 2,6
C	36	52,5 ± 4,9	1,75 ± 0,1	81,81 ± 11,4	26,52 ± 2,2

A. Group A is comprised of test subjects who practice the basic principles of Dr. Mosetter's Glycoplan and received a customized micronutrient formulation based on the HCK modular system.

B. Group B is comprised of test subjects who also changed their nutritional behavior to the Glycoplan, but simultaneously received 1,700 mg of EPA/ DHA in the form of a Norwegian fish oil (company Norsan).

C. The control group (group C) consisted of test subjects who did not receive micronutrients and did not change their nutritional behavior.

As part of the variance analysis (ANOVA) we were able to show that the groups were divided homogenously, and it also shows the individual micronutrients. The intervention phase lasted 24 weeks and the following parameters were captured: micronutrient status, HbA_{1C}, mental state, 24-hour HRV-measurement (stress index, LF/HF-ratio, pNN50).

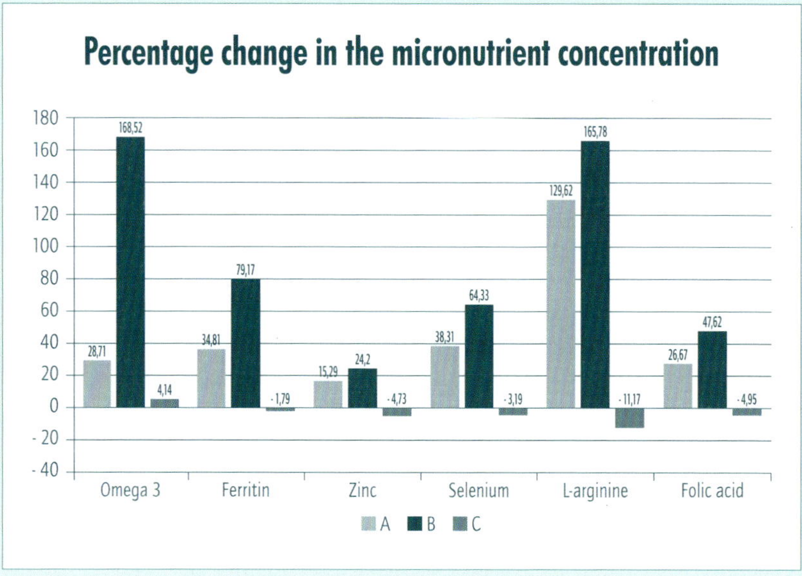

Fig. 41: Percentage change in the micronutrient concentration

Group B shows the biggest changes in the micronutrient status (fig. 41), the HbA1C (long-term glucose levels), as well as disorders. With the aid of 24-hour HRV-measurements it could be proven that the simultaneous intake of the described dosage of Omega-3 fatty acids leads to a significant lowering of the stress index, the vegetative quotient (LF/HF-ratio), and to an increase in the pNN50. This indicates a significantly better balance of the vegetative nervous system.

The targeted intake of Omega-3 fatty acids verifiably improved the absorption of customized micronutrient preparations. But this effect is only possible with the previous dosages of Omega-3 fatty acids of > 1,700 mg EPA/DHA. Further details can be found in the master's thesis (Müller, 2020).

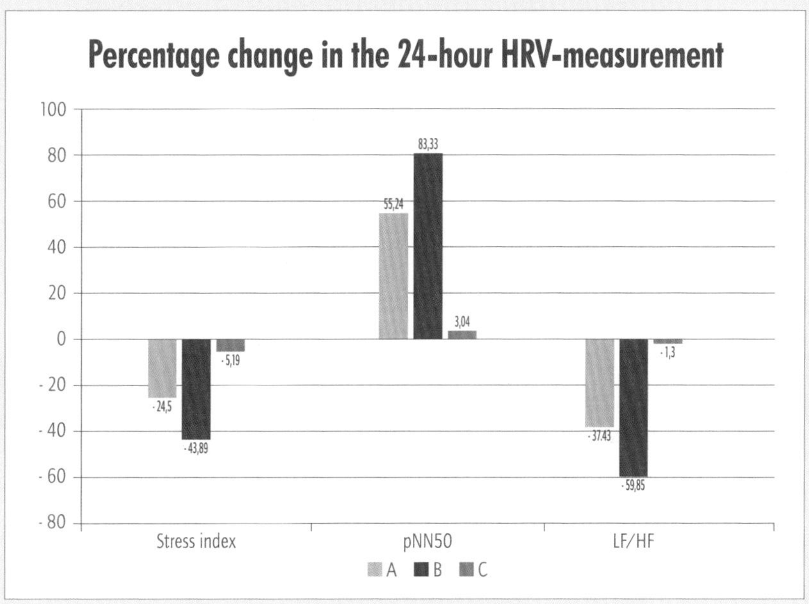

Fig. 42: Percentage change in the 24-hour HRV-measurement

To date there are no comparable studies in the literature. However, at this point we cannot yet provide a detailed answer as to which biochemical effects Omega-3 fatty acids have on the actual change in membrane elasticity. Future additional research projects are recommended to that end.

5.3 Magnesium: The multi-talent among minerals

Magnesium is involved in more than 300 enzymatic processes. These are directly linked to the energy-storage substance ATP (adenosine triphosphate). Magnesium plays a key role in muscle contraction, stimulus conduction in nerve- and muscle cells, in stabilizing cell membranes, and the regulation of heart muscle function. An adequate supply increases stress resistance and prevents premature exhaustion of cellular energy depots and electrolytes.

Magnesium deficiency increases permeability of potassium, which disrupts cellular potassium replenishment and negatively affects physical performance and the heart muscle's stroke frequency. Its antagonistic effect towards potassium protects the myocardial cell from potassium overload, particularly during major exertions, and prevents cardiac arrythmia.

Typical symptoms of magnesium deficiency can be:

- Difficulty staying composed when stressed;
- Muscle weakness;
- Inner restlessness;
- Tendency to painful muscle and calf cramps;
- Poor sleep behavior and verifiably worsening regeneration;
- Training adjustment.

The results from our clinical study show that the magnesium serum concentration increases statistically significantly while the cellular concentration decreases. Hence a valid assertion about the magnesium balance can only be made with the aid of special cellular blood tests that correctly capture the "status quo" of the magnesium balance. More than 85% of tested individuals show definite deficiencies in the cellular magnesium concentration and sleep very restlessly at night.

Better stress tolerance

Mental burdens and stress are a growing societal challenge. People with cellular magnesium deficiencies have significantly higher cortisol levels during stress phases and have trouble unwinding after work and sleep badly. The targeted intake of magnesium resulted in considerably improved stress tolerance and improved sleep behavior. For people with borderline thyroid hormone levels the increased strain on the vegetative nervous system (sympathicotony) can be reduced considerably with a targeted intake.

We conducted a randomized, controlled, two-arm parallel study with 100 test subjects and a study period of more than 90 days (Foundation for Micronutrients, Prevention, Health, Quality of Life, 2015). The main object of the study was the extent to which the mineral magnesium, which, as a natural calcium antagonist, is also very important in cardiology, can impact the sympathovagal balance in combination with endurance training. The effect on the intracellular magnesium concentration was studied as an additional parameter.

We were able to show that the HRV parameters improved significantly in the group that supplemented with 400 mg of magnesium each night one hour before going to sleep: The pNN50 – an indicator for the level of parasympathetic activity – increased. The LF/HF-ratio (vegetative quotient) as well as the stress index – lower values indicate a good balance of the vegetative nervous system – dropped (fig. 43).

The test subjects in the control group showed no positive HRV-parameter changes. Vagus activity and thus the body's ability to adapt and regenerate verifiably increased with the magnesium substitution. The study did not show an effect on the intracellular magnesium concentration.

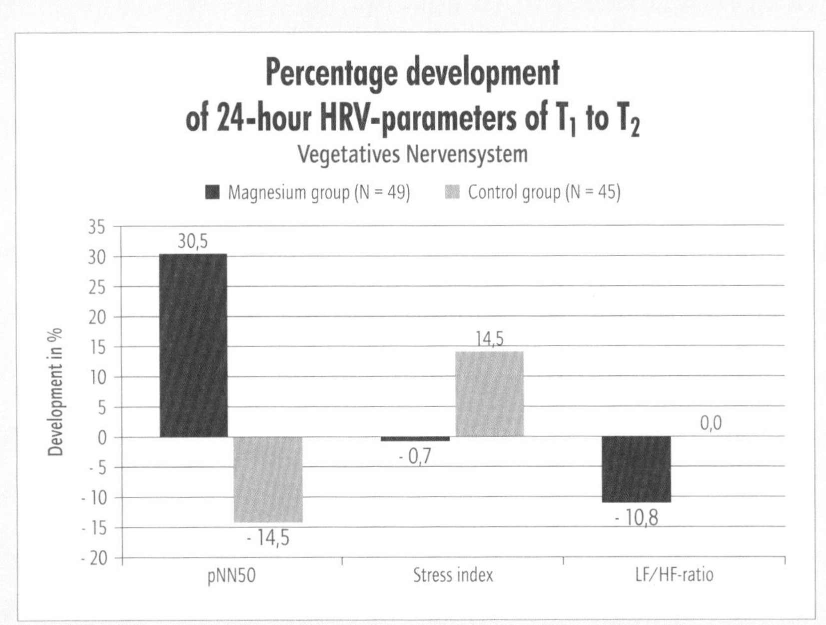

Fig. 43: Percentage development of HRV-parameters

The results of this study indicate that people with mental and physical stresses can benefit from a daily magnesium intake. It not only leads to improved physiological regulation of the sympathetic and parasympathetic efferents, but can also prevent magnesium deficiency and thus additional consecutive symptoms and secondary diseases like, for instance, restlessness, irritability, lack of concentration, insomnia, or depression.

Our initial pre-examinations show that the divided intake of, for instance, magnesium bisglycinate with 200 mg in the morning, afternoon, and evening can have an even greater effect on the vegetative nervous system balance. The next master's project will study this with the aid of a crossover double-blind study.

No optimal endurance development without magnesium

Intracellular magnesium deficiencies in particular can lead to performance losses. About 80% of cellular ATP is bound to magnesium. Magnesium loses its binding partner during exertion-related ATP-consumption. This results in an intracellular release of magnesium and losses in tissue. The shift from intracellular to extracellular space can cause the magnesium serum levels to be deceptively high in spite of a deficiency. The kidneys react to the high magnesium level with increased excretion and thereby promote the magnesium depletion. An optimal cellular magnesium concentration lies 25% above the median values (55 mg/l ery.). However, the current intracellular concentration can only be ascertained with special diagnostics. Cellular deficiencies require a long-term intake of magnesium. A short-term intake of 4-6 weeks cannot replenish cellular stores.

We were able to recognize a direct relationship to an optimal cellular magnesium concentration from the many tests we conducted. During a 3-month training phase, 48 athletes did not receive magnesium preparations while the other group of 52 athletes took 32 mg of magnesium in the morning, afternoon, and evening with food (dosage based on measured magnesium concentration). The ferritin levels of the athletes in both groups that were tested here showed an adequate iron supply at 80.9 ± 8.9.

Athletes with a cellular magnesium concentration of < 44 mg/l ery. grew their endurance performance, measured via the fixed threshold at 4 mmol/l, only by 9.1% (from 3.70 ± 0.30 to 4.07 ± 0.12), while athletes with a magnesium concentration of > 55 mg/l ery. were able to grow the fixed threshold by 14.7% (from 3.81 ± 0.19 to 4.47 ± 0.11).

The training volume was the same for both groups. The significance of this parameter for the development of endurance performance is particularly notable here

5.4 The significance of Vitamin-D

Vitamin-D is a fat-soluble vitamin that is involved in the regulation of the calcium phosphate balance (fig. 44). It is important to callus formation, chondrocyte maturation, and bone mineralization. Vitamin-D has an antithrombotic effect through the activation of thrombomodulin. It lowers blood pressure, has a phagocyte-activating and anti-cancerogenic affect due to inhibiting tumor genesis. It also promotes ß-cell function in the pancreas, increasing insulin secretion!

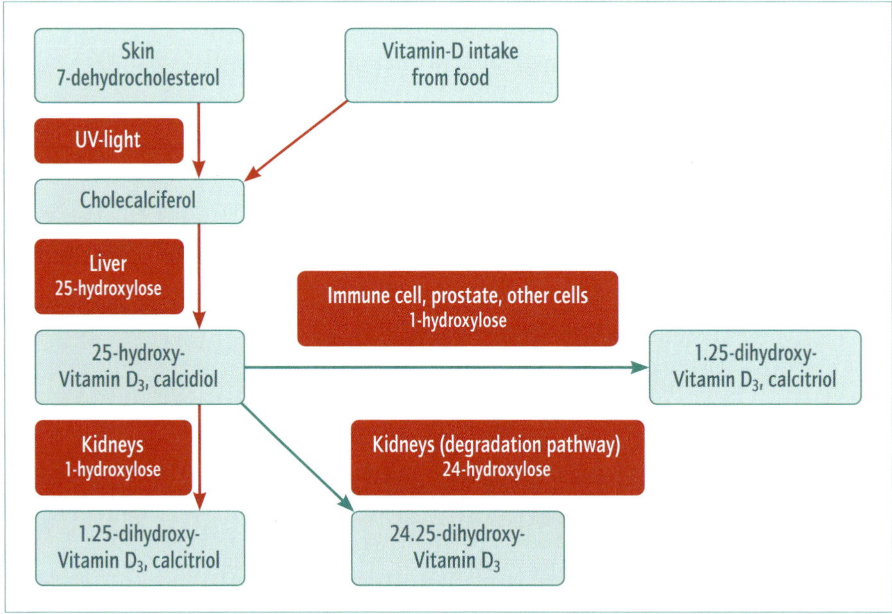

Fig. 44: Vitamin-D metabolism

Which Vitamin-D concentrations are optimal?

A large number of published studies (e.g. Bischoff-Ferrari et al., 2006; Raman et al., 2011) show that Vitamin-D concentrations in the range of approx. 75-125 nmol/l (30-50 µg/l) with respect to 25-OH-D_3 are associated with higher bone density, a lower fracture rate, and decreased incidence of colon carcinoma. Further studies (e.g. published work pertaining to the LURIC-study; Murr et al., 2012) show that Vitamin-D concentrations in this range are also accompanied by fewer incidence of coronary heart disease.

The competence center for complementary medicine diagnostics, Dr. Bayerim synlab MVZ laboratory, was able to show that the increase in 25-OH-D_3, while taking 5000 IU cholecalciferol/per day for 12 weeks varied greatly individually (fig. 45).

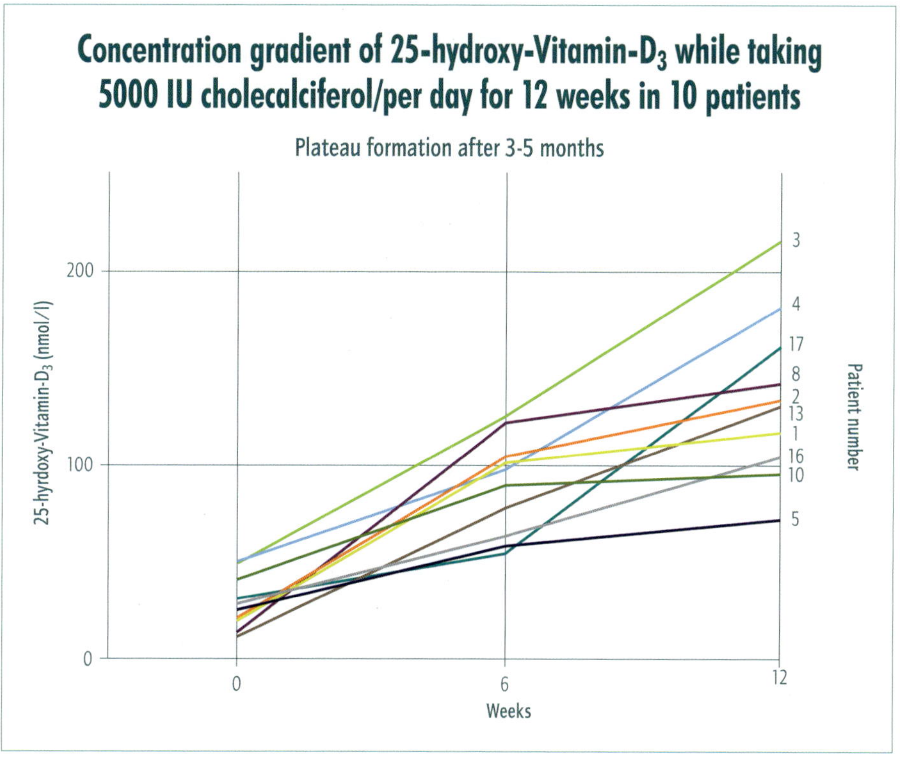

Fig. 45: Increase in 25-Oh-D3 while taking 5000 IU cholecalciferol/per day for 12 weeks in 10 patients

For this reason, no standard dose of Vitamin-D can be determined to achieve optimal serum concentrations. Therefore, what matters is:

Measuring, substituting, monitoring!

6 MILESTONES IN HEALTH CARE MEDICINE

6 MILESTONES IN HEALTH CARE MEDICINE

6.1 The Hydro Cell Key System

The *Hydro-Cell-Key System* (HCK System) is the basis for customized micronutrient formulations and is based on the principle (https://hepart.ch/fileadmin/user_upload/hepart/PDFs/pdf_Produktinformationen_HCK/Hydro_Cell_key_A4_deutsch.pdf):

> *"Everything that grows and serves as food for humans is in a colloidal state. All life on earth is based on colloidal states."*

Galactomannans and polysaccharides in a gel state are carriers of micronutrients that, as a biomatrix, optimize the bioavailability of micronutrients. Micronutrient science also includes knowledge about resorption. It is important to ensure that micronutrients are optimally resorbed with food or in micronutrient supplements.

Another important point is the reciprocal negative impact (antagonism) of many micronutrients among each other, which significantly interferes with the effectiveness of many micronutrient supplements, especially vitamins and minerals, and possibly even triggers manageable and thus undesirable interactions.

A completely new type of embedding of micronutrients in plant-based united cell structures now enables the production of optimally resorbable micronutrient supplements: The process is referred to as the *HCK System* and was created over years of research and developmental work.

We have been using this truly outstanding method for over 15 years to create micronutrient formulations based on the results of our globally unparalleled database.

The HCK carrier substance is the ideal biomatrix: the colloidal system of galactomannans with pseudoplastic flow behavior.

Together with the molecular hydrogen binding forces the galactomannan molecule forms an internal, firmly bound water envelope and a loose *outer water envelope* between the individual galactomannan molecules. The galactomannan form loose junctions among each other, also with built-in micronutrients (fig. 46).

Depending on moisture absorption, the water-soluble galactomannan can be formed as a compact gel (water content up to 97%) all the way to film-forming mediums (water content from 99%). The resulting different exchange surfaces are called on to control the exchange and resorption processes.

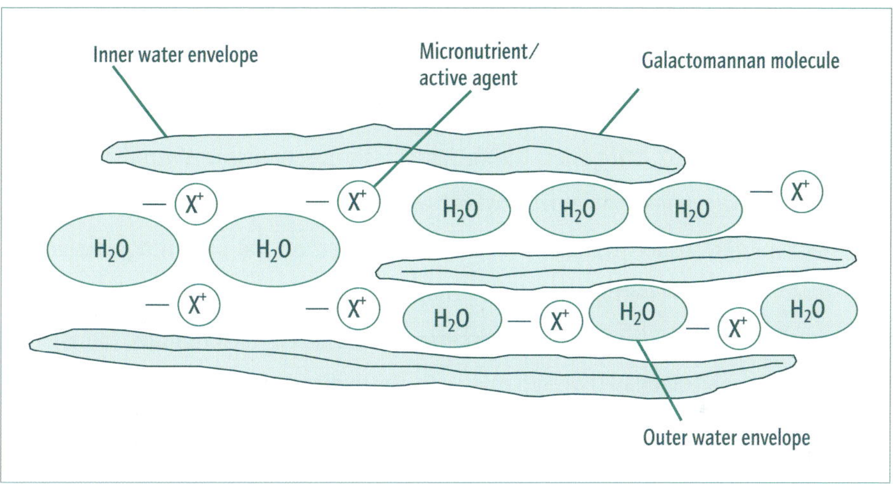

Fig. 46: Hydro-Cell-Key System

One of the most important requirements for micronutrient carrier substances is that they prevent negative effects from micronutrients among each other (antagonism). When absorbing moisture, the micronutrients have their own pH-neutral and non-ionogenic environment within the swollen HCK system. Large exchange surfaces are available for micronutrient access/micronutrient release.

The HCK system prevents the existing reciprocal actions of micronutrients that impede each other. Individual needs for micronutrients can thus be blended and balanced.

6.2 Energy on Prescription

Or: How are customized micronutrient formulations created?

We described the preparation of micronutrient formulations based on the globally unique database in chapter 2.3.2.

There are two methods for the preparation of micronutrient formulations: A blood sample is taken by authorized medical personnel with extensive diagnostics, and secondly the self-test (➡ chapter 6.2.1.3), which enables an individual to obtain a "status quo" of his energy balance without needing assistance to take a blood sample.

The blood-urine values are then entered into the database (10,542 target-/reference values divided into 297 categories) via the anamnesis questionnaire. With the aid of the respective software comparable persons with an identical profile are then identified and their micronutrient concentrations are compared. Based on the percentage deviations from these target-/reference values (➡ chapter 6.2.1.1) and under consideration of the other blood values, a recommendation for a customized micronutrient formulation (small-scale preparation) is then created with the aid of the software, which is then reviewed by a team of experts (physicians, nutritionists, etc.) and finalized. The customer/ patient then chooses which form of the micronutrient formulation he will use.

PLEASE NOTE

The special laboratory facilities measure very accurately. These results are entered into the software of our extensive database and thereby enable the customized creation of micronutrient preparations.

6.2.1 The basis for a customized micronutrient recommendation and formulation

To prepare a customized micronutrient recommendation and preparation we need:

- Age, gender;
- An extensive anamnesis questionnaire;
- Blood and/or urine analyses;
- Calculation of deviations from databank target values.

Blood testing

- **Serum analysis:**
  - This is a routine blood test. It is possible for intracellular deficiencies to have existed for a long time without a change in the serum value (e.g. 1-2% of the entire zinc pool is located in blood serum, 85% is located in the erythrocytes, the red blood cells).
  - **Disadvantage:** No real informational value regarding micronutrient deficiencies.

- **Whole-blood analysis:**
  - Whole-blood analyses provide better informational value because the blood is not centrifuged after the sample is taken and contains both intracellular as well as plasma components.
  - **Disadvantage:** The blood values are subject to short-term dietary influence

- **Intracellular analysis:**
    - Few labs perform intracellular analyses (in erythrocytes). Unfortunately, these labs often don't have adequate target reference values for people with different disorders, diseases, etc.
    - Intracellular analysis is not subject to short-term dietary influence and is critical to the detection of micronutrient deficiencies.

A valid actual state analysis is vital to a successful micronutrient therapy.

Executives, competitive/elite athletes, and people with significant preexisting conditions require a more comprehensive analysis, needed to create the micronutrient formulation (www.saluto.de), and implies more in-depth consulting.

The author and his team founded Energy for Health with the intention of making our results available to a larger population group. Different analysis modules of varying scope and the corresponding price structure are available here.

Basic or extensive tests are advisable depending on the individual, physical state. The results are entered into the globally unique database with the respective software and thus enable the creation of customized micronutrient formulations (www.energyforhealth.de).

Ascertaining the individual energy/micronutrient requirement is based on the following analyses that are based on the requirement profile and current anamnesis:

- The comprehensive anamnesis questionnaire enables classification into the cluster of the 10,542 target/reference values divided into 297 categories.
- Full blood panel (thyroid hormones, liver/kidney values, HbA_{1C} value, CRP value).
- 1-FABP-parameter to preclude current intestinal barrier dysfunction;
- Special cellular micronutrient analyses;
- Functional energy metabolism;
- Concentration of various amino acids;

- Use of endogenous structural proteins in the energy metabolism (pyridinium crosslinks).
- 24-/48-hour HRV-measurement;
- Daily cortisol profile (saliva).

6.2.1.1 Brief description of individual blood, urine, and saliva analyses

Functional energy metabolism

The citric acid cycle forms the central switch point of the entire metabolism (➡ chapter 2.3). The measurement is performed via analysis of the morning's first urine. In recent years, the author and his team had more than 7,500 tests carried out and have been able to complete group-specific classification of the results into specific clusters.

In the citrate cycle seven different acids and metabolites are measured that, based on our results, point to impaired activity of important enzymes in the energy metabolism:

- Citric acid
- Cis-aconitic acid
- Alpha-ketoglutaric acid
- Succinic acid
- Fumaric acid
- Malic acid
- Pyruvate

The targeted intake of customized micronutrient formulations leads to economization of the energy metabolism. The example shown (fig. 47) illustrates that prior to micronutrient therapy some acids and metabolites are in the red and indicate that the functional energy metabolism is significantly impaired (blue columns). After 12 weeks of a customized micronutrient intake the energy metabolism has been economized and now registers and optimal energy metabolism.

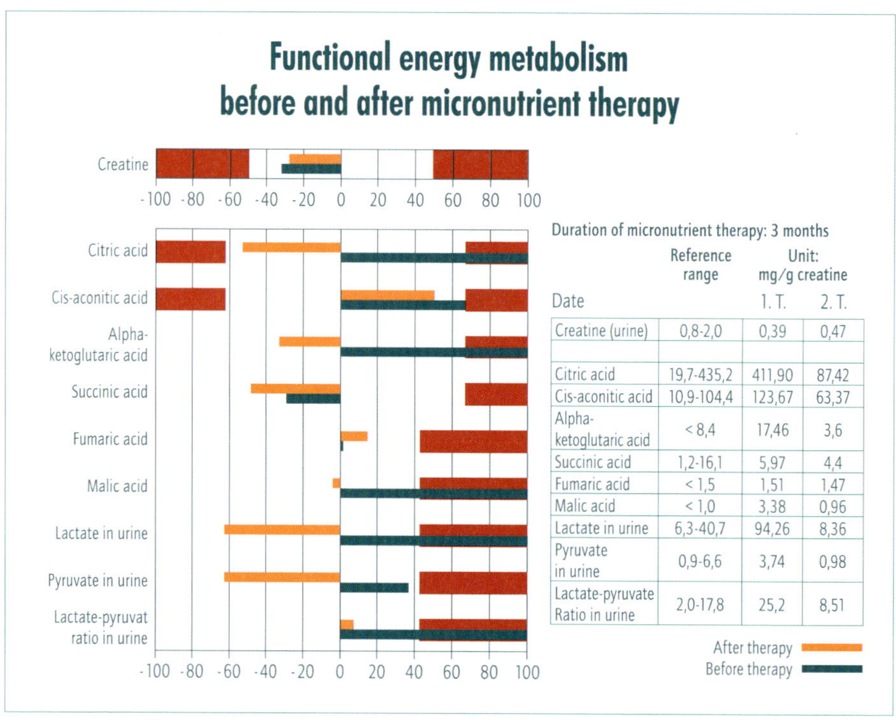

Date	Reference range	Unit: mg/g creatine 1. T.	2. T.
Creatine (urine)	0,8-2,0	0,39	0,47
Citric acid	19,7-435,2	411,90	87,42
Cis-aconitic acid	10,9-104,4	123,67	63,37
Alpha-ketoglutaric acid	< 8,4	17,46	3,6
Succinic acid	1,2-16,1	5,97	4,4
Fumaric acid	< 1,5	1,51	1,47
Malic acid	< 1,0	3,38	0,96
Lactate in urine	6,3-40,7	94,26	8,36
Pyruvate in urine	0,9-6,6	3,74	0,98
Lactate-pyruvate Ratio in urine	2,0-17,8	25,2	8,51

Fig. 47: Functional energy metabolism before and after micronutrient therapy

Indicator for cartilage and bone metabolism (Pyridinium crosslinks)

Bone but also cartilage consists of collagen molecules that are stabilized by crosslinks. These crosslinks are

- Primarily deoxypyridinoline (DPD) in bone and
- Pyridinoline (PD) in cartilage.

During amplified degradation processes these products of crosslinking are released into the blood and afterwards discharged in urine. The amount of the discharged pyridinium crosslinks depends on the volume of the degradation processes. Since the discharges are not impacted by the neosynthesis of bone nor the collagen-containing food components, deoxypyridinoline is considered the currently best marker for selective assessment of bone resorption.

Based on our experiences in measuring pyridinium crosslinks in 14,400 elite athletes and 6,350 recreational athletes, this parameter is a good indicator for the extent to which the athlete's energy metabolism has used endogenous structural proteins or rather the cartilage and bone metabolism that are of fundamental importance to the preservation of function of many connective tissue structures (ligaments, tendons, cartilage). Many athletes show definite micronutrient deficiencies that verifiably block enzymes during energy production and cause the organism to resort to using endogenous structural proteins in the short-term.

PD and/or DPD levels that are clearly above the normal range show that energy production in athletes takes place short-term via endogenous structural proteins. Optimization of the micronutrient balance is urgently needed here. Long-term elevated levels can verifiably increase risk of injury.

Pyridium crosslinks (ascertained from the morning's second urine) can reveal how extensively the energy metabolism uses endogenous structural proteins. A PD-level of >85 nmol/mmol creatinine and DPD-levels of > 15nmol/mmol creatinine can have a negative effect on mental and physical performance and increase the risk of injury.

Fig. 48 shows a definite reduction in pyridinium crosslinks with a targeted micronutrient intake in 144 marathon runners (ages 27.1 ± 3.1) with a performance level of 2:30 to 2:50 h/42.195 km over a time period of eight months. It is evidence of the decreased use of structural proteins for energy production.

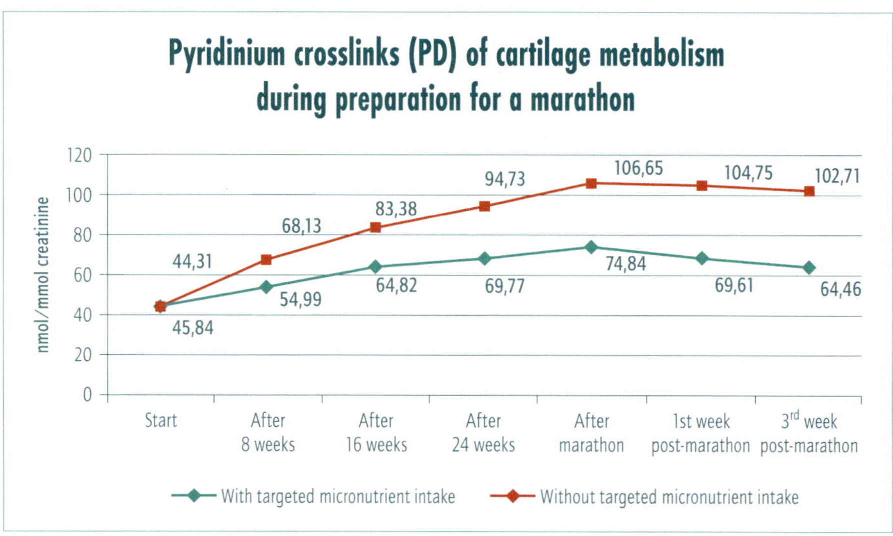

Fig. 48: Pyridinium crosslinks of cartilage metabolism during preparation for a marathon

I-FABP measurement to test for current intestinal barrier disorder

I-FABP (➡ chapter 3.2) is released into the circulation when the intestinal epithelium is damaged, and can be measured in serum. Compared to Zonulin, this parameter has more informational value.

Analyses of different amino acid concentrations

Mental and physical stress can significantly increase demand. Amino acids as nutrients have a direct impact on your personal wellbeing (➡ chapter 5.1). You can influence your status yourself and achieve a "good mood" by choosing a proper diet.

Euphoria and optimism are a reflection of your amino acids. An optimal supply with the basic amino acids listed in the appendix quickly makes sadness and anxiety fade away. But an immune-system-stabilizing and function-preserving effect of the stressed connective-tissue structures above 20% of the respective 10,543 target/reference values divided into 297 categories is also evident.

Analyses of different micronutrient concentrations

The database contains results from 32,500 whole-blood analyses and intraerythrocytic analyses of individual micronutrients, enabling evaluation of the individual micronutrient concentrations in the respective group of people based on age, gender, condition, preexisting conditions, and athletic activity. With intravenous blood sampling the analyses of intraerythrocytic micronutrient concentrations include the following parameters: magnesium, zinc, selenium, Vitamin-B_1, B_2, B_6, B_9; HS-Omega-3 index as well as in serum: holoTC (active Vitamin-B_{12}) and Vitamin-D.

The self-test with the TAP device (without intravenous blood sampling); ➡ chapter 6.2.1.3) includes TSH-basal value, HbA1C-value, ferritin, Vitamin-D.

No activity restrictions in the entire metabolism can be seen above 20% of the respective target value of these micronutrients.

Daily cortisol profile

The daily cortisol profile is a good alternative to the 24-/48-hour HRV measurement for people with cardiac arrythmias.

6.2.1.2 Sample analysis of a test subject

Analyses and evaluation of amino acid concentrations

The following example shows in detail how the analysis functions. The test subject is a 42-year-old woman with many disorders.

The database software uses algorithms to select individuals with a similar profile based on gender, age, disorders, preexisting conditions, athletic activity, etc., and establishes the percentage deviations of the test subject's median values relative to these individuals (zero level in illustration).

Tab. 5-8 show the laboratory's general reference ranges and measuring results for individual amino acids. The deviations of these measuring results from the respective median values of comparable groups of people (here 450 women) are then calculated based on gender, age, preexisting conditions, disorders, athletic activity from the database and the software. In fig. 49-52 these results have been diagrammed as the deviation of the test subject's measured values from the control group's median.

Tab. 5: General reference ranges and measuring results of individual amino acids. Function-preserving tasks of stressed connective-tissue structures of the active/passive locomotor system and stabilization of the immune system.

	Reference ranges of laboratory (in mg/dl)	Measured value of 42-year-old test subject (in mg/dl)	Median value of 450 comparable women based on age, disorders, preexisting conditions, athletic activity
Arginine	0,697-2,939	1,536	2,145
Glutamine	5,188-11,692	7,267	9,015
Glutamic acid	0,147-4,725	0,195	3,812
Glycine	1,239-3,566	2,721	2,613
Methionine	0,149-0,797	0,280	0,621
Proline	0,978-5,742	1,478	4,123

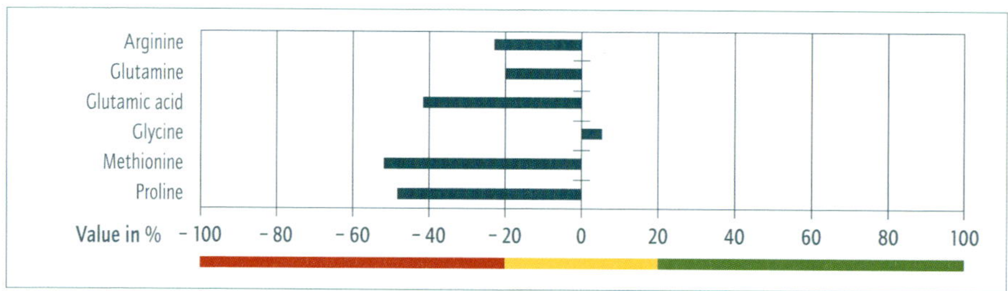

Fig. 49: Deviations of individual amino acids. Function-preserving tasks of stressed connective-tissue structures of the active and passive locomotor system as well as stabilization of the immune system in the test subject from the 0 level to the median of the comparison group

Tab. 6: General reference ranges and measuring results of individual amino acids of the energy metabolism and for transmitter activation

	Laboratory reference ranges (in mg/dl)	Measured value of a 42-year-old female test subject (in mg/dl)	Median value of 450 comparable women based on age, disorders, preexisting conditions, athletic activity
Alanine	2,317-5,479	3,849	4,213
Aspartic acid	< 0,333	0,002	0,297
Histidine	0,76-1,707	1,161	1,421
Lysine	1,827-4,385	1,974	3,213
Ornithine	0,595-2,049	0,912	1,421
Taurine	0,501-3,379	0,724	3,012
Tyrosine	0,698-1,828	0,887	1,523

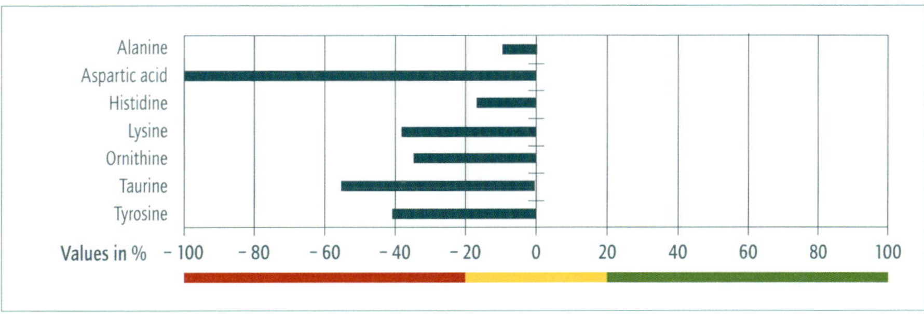

Fig. 50: Deviation of individual energy-metabolism amino acids and for transmitter activation of the test subject from the zero level to the comparison group's median

Tab. 7: General reference ranges and measuring results of individual amino acids in the energy metabolism and glucose balance

	Laboratory reference ranges (in mg/dl)	Measured value of a 42-year-old female test subject (in mg/dl)	Median value of 450 comparable women based on age, disorders, preexisting conditions, athletic activity
Leucine	1,115-3,624	1,254	3,122
Isoleucine	0,525–2,143	0,671	1,624
Valine	1,640-4,517	1,890	3,816

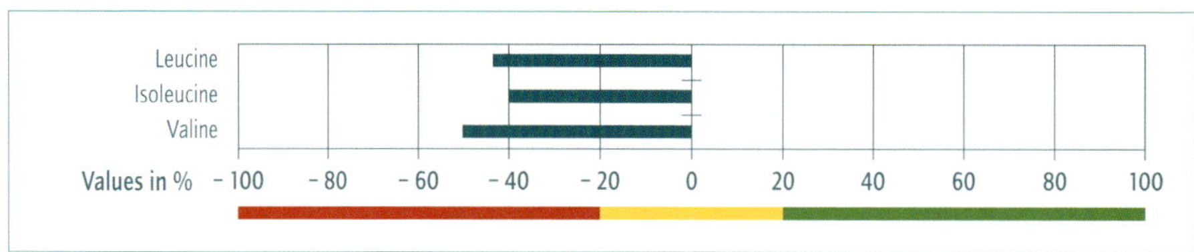

Fig. 51: Deviation of the test subject's individual energy-metabolism and glucose-balance amino acids from the zero level to the comparison group's median

Tab. 8: General reference ranges and measuring results of phenylalanine and tryptophan, precursor substances for calming and activating neurotransmitters in the brain metabolism

	Laboratory reference ranges (in mg/dl)	Measured value of a 42-year-old female test subject (in mg/dl)	Median value of 450 comparable women based on age, disorders, preexisting conditions, athletic activity
Phenylalanine	0,743-2,387	0,690	2,156
Tryptophan	0,715-1,829	1,175	1,721

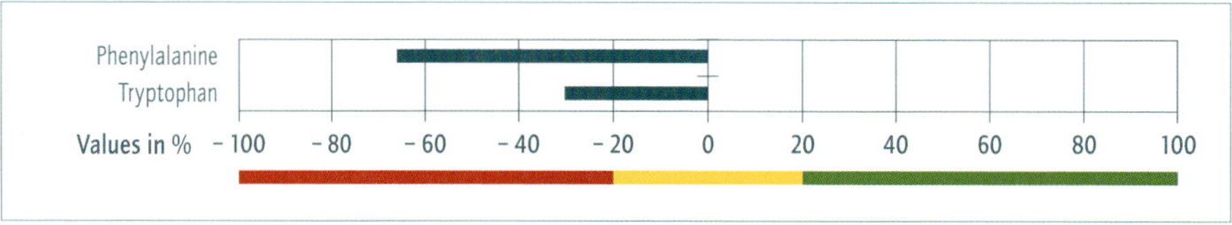

Fig. 52: Deviation of phenylalanine and tryptophan, precursor substances for calming and activating neurotransmitters in the test subject's brain metabolism from the zero level to the comparison group's median

Analyses of the different micronutrient concentrations in different blood compartments

Tab. 9 shows the general reference ranges and the test subject's measuring results with respect to individual micronutrients. The deviations of these measuring results from the respective median values of comparable groups of people (here 450 women) are then calculated from the database and software based on gender, age, preexisting conditions, disorders, and athletic activity. These results have been diagrammed in fig. 53 as the deviation of the test subject's measuring results from the control group's median.

Tab. 9: General reference ranges and measuring results of micronutrient concentrations

	Laboratory reference ranges	Unit	Measured value of a 42-year-old female test subject	Median value of 450 comparable women based on age, disorders, preexisting conditions, athletic activity
Magnesium ery.	40,1-64,4	mg/l	36,1	46,3
Zinc ery.	8,8-16,0	mg/l	9,8	9,8
Selenium ery.	116-356	µg/l	112	197,0
Ferritin, women	13-150	ng/ml	40,1	48,1
Vitamin B_1 ery.	28-85	µg/l	54,0	74,3
Vitamin B_2 ery.	90-350	µg/l	291,0	349,1
Vitamin B_6 ery.	8,7-27,2	µg/l	41,2	257,6
Folic acid (B_9) ery.	212-534	µg/l	226	257,4
HoloTC (active B_{12})[a]	> 75	pmol/l	72,0	110
Vitamin D[b]	75-125 > 30-50	nmol/l µg/l	61,00	112
HS-Omega-3 Index (ery.)	8-11	%	5,83	8,0
Total protein	6,6-8,7	g/dl	6,9	7,5

[a] Target range: >130 pmol/l [b] Target range: > 130 and < 180 nmol/l

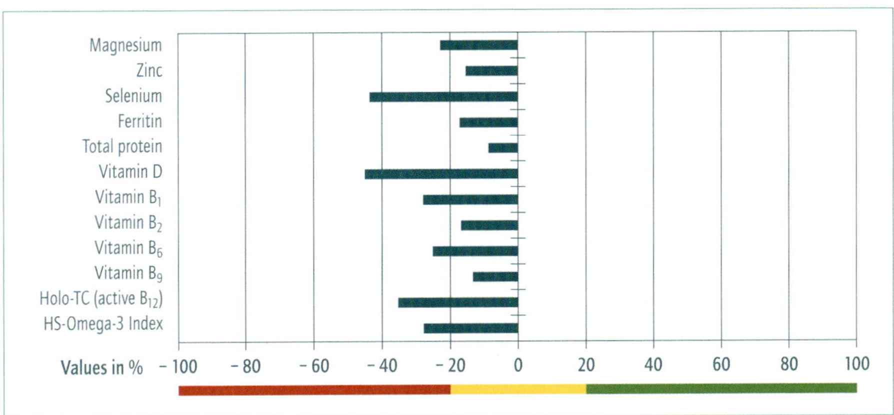

Fig. 53: Deviation of the test subject's individual micronutrient concentrations from the zero level to the comparison group's median

Analysis of sense of wellbeing with the 24-/48-hour HRV-measurement

The 24- or 48-hour HRV measurement helps to capture the physical state of an individual easily and without complication (➡ chapter 2.2.1).

The example of the 42-year-old woman with various disorders illustrates the changes in the vegetative nervous system due to a customized micronutrient formulation.

With the customized micronutrient intake, the stress index (fig. 54) dropped or rather normalized significantly as compared to the first test.

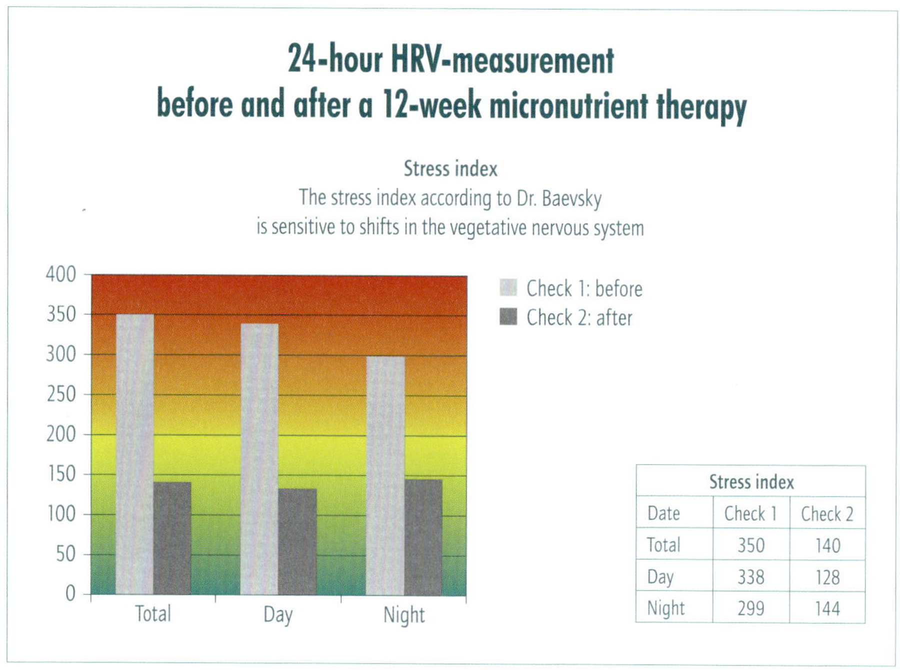

Stress index		
Date	Check 1	Check 2
Total	350	140
Day	338	128
Night	299	144

Fig. 54: 24-hour HRV measurement before and 12 weeks after micronutrient therapy: stress index

At the beginning of the study, the pNN50 as the indicator of "vegetative recovery" (activation of the parasympathetic nervous system) was very high > 30, and indicates absolute chronic fatigue syndrome (reference: 10-25). After the targeted micronutrient intake, the pNN50 only shows values between 12 and 14 during the day and a very positive effect on the vegetative nervous system (fig. 55).

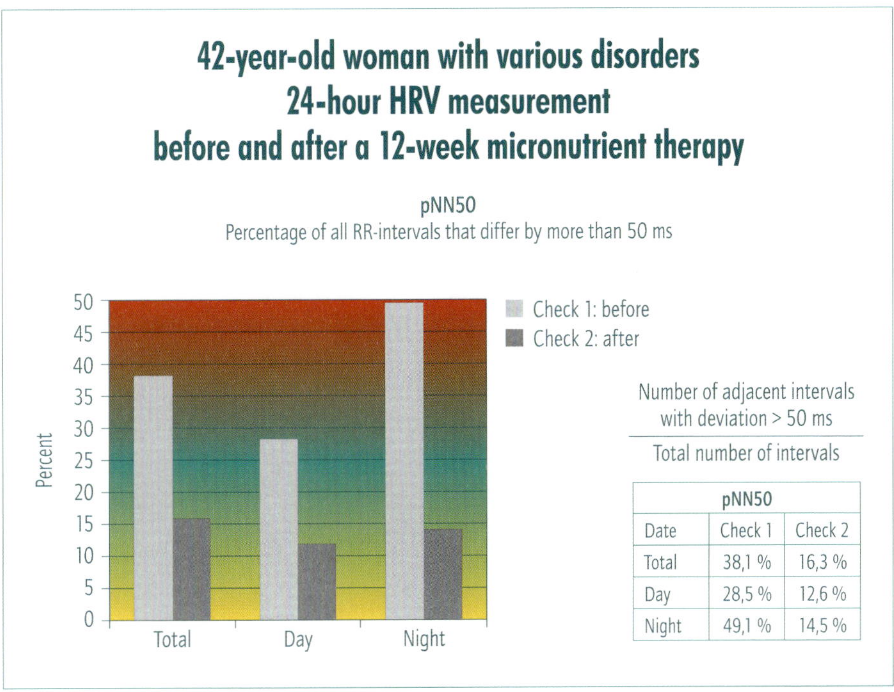

pNN50		
Date	Check 1	Check 2
Total	38,1 %	16,3 %
Day	28,5 %	12,6 %
Night	49,1 %	14,5 %

Fig. 55: 24-hour HRV measurement before and after 12 weeks of micronutrient therapy: pNN50

Analog the pNN50, the reduction of the LF/HF-ratio (vegetative quotient) of between 2.0 and 3.0 due to the targeted micronutrient intake shows a very good development and indicates the extraordinary impact an intake of lacking micronutrients can have on a person's wellbeing (fig. 56).

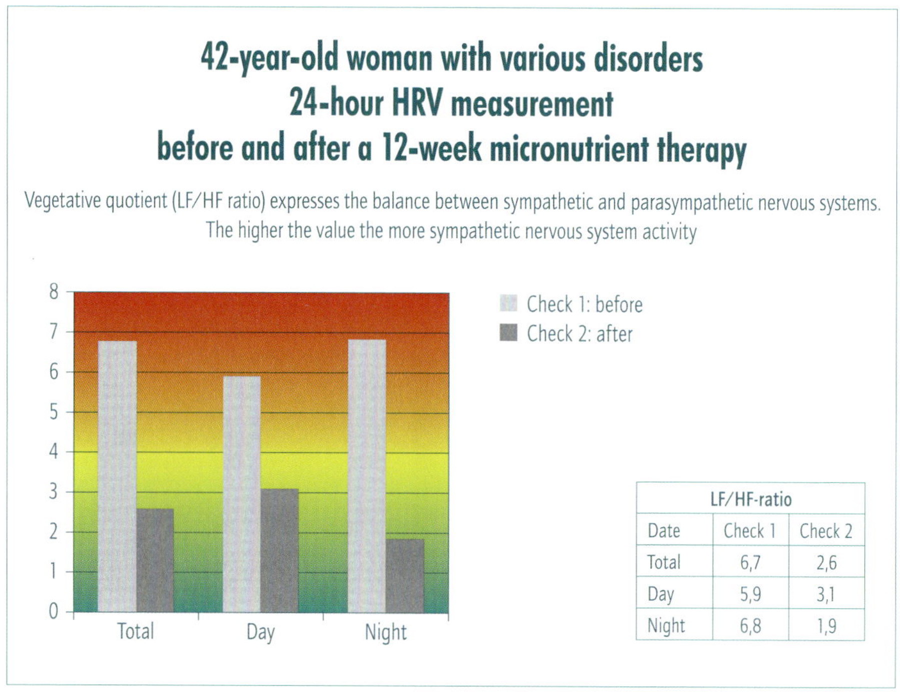

LF/HF-ratio		
Date	Check 1	Check 2
Total	6,7	2,6
Day	5,9	3,1
Night	6,8	1,9

Fig. 56: 24-hour HRV measurement before and 12 weeks after micronutrient therapy: vegetative quotient

Fig. 57 shows the positive effect micronutrient therapy had on the wellbeing of the 42-year-old woman.

42-year-old woman with various disorders
24-hour HRV measurement
before and after a 12-week micronutrient therapy

Before

- Absolute chronic fatigue syndrome
- Constant inner restlessness
- Frequent trouble concentrating
- Permanent muscle tension
- Trouble falling asleep/sleeping through the night
- Very stressed vegetative nervous system
- No longer able to perform athletic training

After

- Good mental and physical performance capacity
- Rarely in pain
- Discontinued all medications
- Easily falls asleep/sleeps through the night
- Is able to resume athletic training without any problems

Fig. 57: Subjective sense of wellbeing before and after 12 weeks of micronutrient therapy

Daily cortisol profile

The result of the daily cortisol profile of the 42-year-old woman with various disorders (fig. 58) documents the absolute chronic fatigue she described. Just half an hour after getting up in the morning (awakening response), she already shows a cortisol level of 0.199 µg/l, which is consistent with absolute exhaustion (➡ chapter 2.2.2).

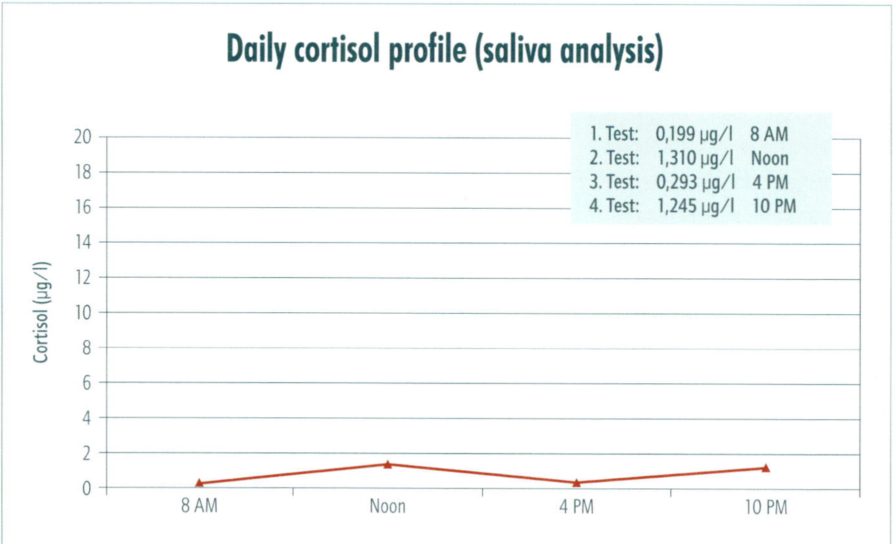

Fig. 58: Daily cortisol profile (saliva analysis)

6.2.1.3 TAP self-test: An alternative to intravenous blood sampling

This self-test is an alternative for the creation of customized micronutrient preparations. Previously venipuncture and finger-prick sampling were "points of criticism" in our diagnostic investigations, which is why we needed a simple, safe, standardized and pain-free method for blood-sampling within and outside the traditional healthcare facilities.

This interest does not just apply to day-to-day operations in clinics and hospitals. It also applies to all jobs with additional qualifications. The number of jobs requiring experience in the area of metabolism is growing, which results in a greater demand for lab diagnostics in therapeutic professions that perform diagnostic analyses.

More recently, the capillary blood collection device TAP (Touch-activated phlebotomy) was introduced as innovative and user-friendly, for quick handling and pain-free collection of small blood samples for clinical analysis (Blicharz et al., 2018). It is argued that this sterile single-use device is superior to conventional blood-sampling via venipuncture or other methods like the finger-prick because the donor is able to complete the process pain-free and it can be easily used by untrained personnel.

The Foundation for Micronutrients, Prevention, Health, and Quality of Life together with a graduate from the master and certificate course in micronutrient therapy and regulatory medicine have now examined the validity the TAP device and the exact relationship between whole-blood analysis and cellular analysis in the TAP device as part of a SIP project. In the meantime, more studies with a similar question have been completed to be able to confirm the primarily ascertained results.

> The goal of the SIP project was to verify the consistency of results from whole-blood analyses with results from analyses with washed erythrocytes. Is a minimally invasive blood test a reliable alternative to traditional, standardized venipuncture?

The comparison results between the whole-blood analysis with TAP and intravenous (IV) analysis with respect to micronutrient concentrations show scientifically valid correlations (tab. 10, fig. 59 and fig. 60).

Tab. 10: Comparison of magnesium, zinc, selenium ascertainment in whole-blood versus erythrocytes.

Whole blood vs. erythrocytes					
Parameter	N	Whole blood vs. erythrocytes	M	SD	R (p)
Mg (mmol/l)	25	Whole blood Erythrocytes	1,49 0,71	0,119 0,088	,442 (p - 0,027)
Zn (mmol/l)	25	Whole blood Erythrocytes	0,09 0,06	0,014 0,012	,826 (p < 0,001)
Se (µmol/l)	25	Whole blood Erythrocytes	1,54 0,62	0,278 0,168	,821 (p < 0,001)

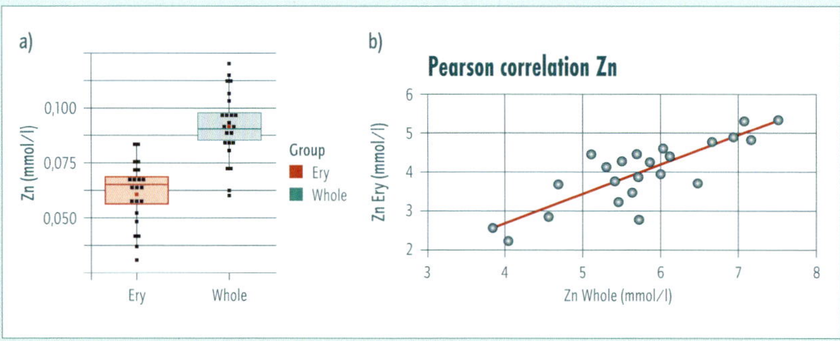

Fig. 59: Quantification of zinc samples in erythrocytes and whole blood

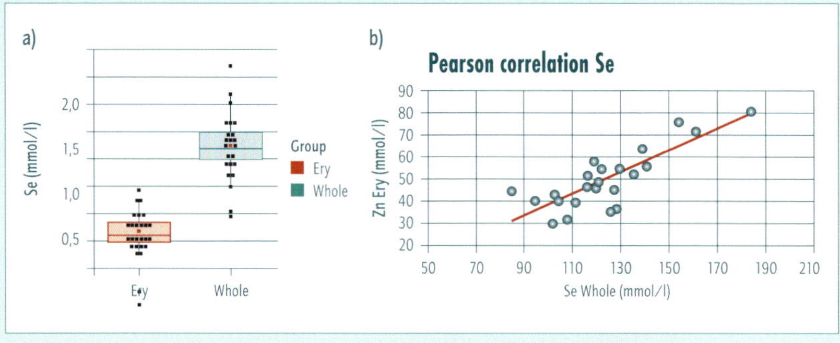

Fig. 60: Quantification of selenium probes in erythrocytes and whole blood

Our long-time results from 32,500 whole-blood analyses and intraerythrocytic analyses via IV-blood sampling show correlations in magnesium of 0.862, in zinc 0.893, in selenium 0.871, Vitamin B1 0.899, and thereby document a correlation. Further details about the informational value of whole-blood analyses vs. intraerythrocytic analyses can be found in chapter 2.3.1.

PLEASE NOTE

There is a scientifically verifiable correlation between the two blood compartments.

To this question, 30 test subjects, average age 41.5, had blood drawn with the TAP device and via the conventional method. Parameters measured were ferritin, TSH, HbA_{1C}, Vitamin D, zinc, selenium, and magnesium, and the measuring results were studied for correlations. The following hypothesis was tested:

- H1: There are positive relative and absolute correlations between intravenous and TAP device ferritin values.
- H2: There are positive relative and absolute correlations between intravenous and TAP device TSH values.
- H3: There are positive relative and absolute correlations between intravenous and TAP device HbA1C values.
- H4: There are positive relative and absolute correlations between intravenous and TAP device Vitamin D values.
- H5: There are positive relative and absolute correlations between intravenous and TAP device zinc values.

- H6: There are positive relative and absolute correlations between intravenous and TAP device selenium values.
- H7: There are positive relative and absolute correlations between intravenous and TAP device magnesium values.

With the test subjects' wide age variance, positive results such as ferritin, HbA_{1C}, TSH, and Vitamin D, in spite of a low sample number, suggest high general informational value with respect to their individual values (tab. 11).

Tab. 11: Overview of analysis results from TAP vs. IV blood sampling

TAP vs. IV blood								
Parameter	N	w/m	Age	TAP vs. IV	M	SD	ICC (CI)	R (p)
Ferritin	29	20/9	46,6 (15,8)	TAP blood IV blood	80,2 79,7	47,7 48	,978 (,953-,990)	,956** ($p < 0,0001$)
TSH basal (µU/L)	28	19/9	46,6 (15,8)	TAP blood IV blood	1,85 1,84	0,93 0,85	,845 (,664-,929)	,728** ($p < 0,0001$)
HbA_{1c} (%)	29	20/9	46,6 (15,8)	TAP blood IV blood	5,1 5,3	0,3 0,32	,900 (-,050-,979)	,987** ($p < 0,0001$)
Vitamin D (nmol/l)	27	20/7	46,6 (15,8)	TAP blood IV blood	82,8 76,9	64,99 63,58	,994 (,975-998	,991** ($p < 0,0001$)
Zn (mmol/l)	29	20/9	46,6 (15,8)	TAP blood IV blood	6,86 6,26	1,47 0,81	,648 (,246-835)	,626** ($p < 0,0001$)
Se (µg/l)	29	20/9	46,6 (15,8)	TAP blood IV blood	113,5 131,1	21,21 19,44	,621 (-,060-,848)	,609** ($p < 0,0001$)
Mg (mmol/l)	29	20/9	46,6 (15,8)	TAP blood IV blood	1,59 1,49	0,2 0,16	,765 (,318-,904)	,722** ($p < 0,0001$)

TAP = Touch-activated Phlebotomy, IV = intravenous, N = total, f/m = female/male, M = median value, SD = standard deviation, ICC = interrater correlation coefficient, CI = confidence interval, R = correlation coefficient, p = significance, TSH = thyroid-stimulating hormone, µU = micro units, Mg = magnesium, Zn = zinc, Se = selenium, ** correlation is significant at the 0.01 level.

The results show that significant correlations can be observed for hypotheses H1-H4. The extent of the correlation indicates a certain variability. Correlations are also verifiable for the minerals of magnesium, zinc, and selenium, but the variances are significantly greater.

We therefore currently still recommend not measuring minerals in whole blood with the TAP device since the variation range is too great. The IV analysis is on average about 0.1-0.2% higher for HbA_{1C} values as compared to the TAP-device analysis.

Overall, the study results suggest that the self-test TAP device allows healthcare professionals to receive scientifically valid blood results in addition to the anamnesis data, that can be integrated into the therapy concept. Since the device is easy to use and does not require handling by trained medical staff, it can also make drawing and analyzing blood at home via the professional health sector possible. However, clients would need to be made explicitly aware of their self-responsibility to minimize time-delayed measuring errors.

The blood analysis of TSH-basal values, ferritin, HbA_{1C}, and Vitamin-D concentration already provide important findings for further procedures. With the aid of the database combined with the analysis of the functional energy metabolism (first urine of the morning), which is the metabolism's central switch point (➡ chapter 2.3), a very good micronutrient recommendation can be made.

Our recommendation: Foregoing the measuring of minerals in whole blood with the TAP device because the range of variation is too great and thus cannot be assigned to the individual clusters such as gender, disorders, preexisting conditions, athletic activity.

6.3 Customized micronutrient formulations: What do they look like in practice?

The basic formulation takes place via the previously described Hydro Cell Key System by HCK (➡ chapter 6.1). The customer/patient is given a prescription formulation that has been compounded by a pharmacist based on the test results and the globally unique "evidence-based database and software". The individual micronutrients are incorporated into the plant matrix. Each formulation contains many plant extracts such as citrus flavonoids all the way to green tea extracts. As an alternative the dosage with a therapeutic effect can also be created from a combination of many mono and combination preparations combined with secondary vital substances. Most important here is the quality of the micronutrients. Easy handling is recommended for long-term therapy compliance.

In some cases, an additional targeted daily intake of several non-preparation and amino acid blends can be necessary. This would be primarily special collagen peptides which, depending on the medical findings, contain L-tryptophan or no L-tryptophan. In some cases, a short-term higher intake of Vitamin D, Vitamin C, or other micronutrients is also recommended, beneficial, and very effective in the treatment of many underlying diseases.

The Omega-3 fatty acids are administered via a fish oil produced in Norway or alternatively via a vegan algae oil. Here the dosage is of particular importance to achieve verifiable effects. The targeted and isolated intake of magnesium, for instance as magnesium bisglycinate in a 100 mg capsule spread out over the day, can help activate the parasympathetic nervous system and verifiably improve regulation of the vegetative nervous system.

But the customized micronutrient formulation is only optimally effective in combination with the basic principles of Dr. Mosetter's "traffic light system" (➡ chapter 4.7.1). But in some cases, infusion solutions of individual micronutrients can also be a beneficial addition to acute therapy under medical supervision.

Reference ranges of a custom micronutrient formulation as a basic provision

Agent	Daily dose	Agent	Daily dose
Vitamins		**Trace elements**	
Vitamin A (retinol)	0,5-1,0 mg	Chrome	50-500 µg
Vitamin B_1 (thiamine)	10-70 mg	Selenium	50-200 µg
Vitamin B_2 (riboflavin)	10-70 mg	Zinc	10-52 mg
Niacin (Vitamin B_3)	10-70 mg		
Vitamin B_6 (pyridoxin)	10-60 mg	**Minerals**	
Vitamin B_{12} (cyanocobalamin)	30-2.000 µg	Silicon	20-40 mg
Vitamin C (ascorbic acid)	500-3.000 mg		
Vitamin D	25-200 µg	**Quasi-vitamins**	
Natural Vitamin E	75-250 mg	Choline	40-100 mg
Of that alpha tocopherol	50-180 mg	Coenzyme Q_{10}	30-120 mg
Gama tocopherol	4-21 mg	Inositol	20-100 mg
Natural carotenoids	4-10 mg	L-carnitine	100-1.000 mg
Of that alpha carotin	30-70 µg	PABA	40-60 mg
Beta carotin	1,0-1,9 mg		
Cryptoxanthin	10-15 µg	**Plant extracts**	
Lutein	3-9 mg	Citrus flavonoids	
Zeaxanthin	11-15 mg	Green tea extracts	
Biotin (Vitamin H)	15-50 µg	Red wine extract	
Folic acid (Vitamin B_9)	600-1.600 µg		
Pantothenic acid	20-80 mg	**Fiber**	
Additional intake of individual amino acids as needed		Guar gum	
		Inulin	
		HPM cellulose	
Phenyl alarin	500-1.500 mg		
Arginine	3.000 mg		
Tyrosine	60-400 mg		
5-HTP	50-300 mg		
Glutamine	500-3.000 mg		

Daily intake **15-40 ml**
Morning and afternoon each 10-40 ml
After three months only half of the daily quantity
If necessary additional
10 – 30 g of collagen peptides with or without L-tryptophan
- Jodid 50-300 µg
- Magnesium 100 mg capsule as magnesium bisglycinate. 1 - 6 capsules over the course of the day.
- 2 Tbs. Norsan Omega-3 oil (1 Tbs. = 1700 mg EPA/DHA), 1/1/0 after eight weeks only in the morning.
- Alternatively vegan Norsan algae oil: 1-3 tsp. (1700 EPA/DHA – 5100 EPA/DHA)
- Evenings 1 hour before bed 5-HTP (hydroxytryptophan) 50 – 300 mg

Arginine in powder form 1 – 7 g

Fig. 61: Reference ranges of customized micronutrient formulations as a basic provision

6.3.1 Attention deficit/hyperactivity disorder (ADHD)

A combination of symptoms such as motor restlessness, unusual behavior patterns, and difficulty concentrating was first described by Melchior Adam Weikard in 1775. Attention deficit/hyperactivity disorder (ADHA) is one of the most common mental disorders in children and adolescents.

ADHD has three primary symptoms: attention deficit and/or impulsivity as well as hyperactivity. To diagnose ADHD, the primary symptoms must be outside of the normal range of a child's age and stage of development, they must occur trans-situationally and be accompanied by limitations to social, academic, or professional functional performance.

There are many studies and thoughts about the causes, triggers, and types of treatments, but the proper care for those afflicted so they may participate in life without limitations is frequently still not within reach.

The typical disorders "attention deficit and hyperactivity" can be induced by a micronutrient deficiency. It was the goal of a master's thesis to show that the participation of affected ADHD children could be improved with micronutrient therapy (also see: *Stellenwert der Mikronährstoffe bei Kindern mit einer Aufmerksamkeits-/ Hyperaktivitätsstörung (Micronutrient status in children with ADHD.* FHM publication series, 12th edition, pg. 60-71).

6.3.1.1 ADHD – retrospective evaluation of micronutrient diagnostics and therapy

Methodology

The study used data from 162 children between the ages of seven and 12 who had been diagnosed with ADHD. The retrospective data analysis was done on the empirical data of the test subjects, including state of health, blood and urine analysis, and micronutrient diagnostics, which were provided by SALUTO.

Reference ranges of customized micronutrient formulations for ADHD children with the aid of the HCK system

Agent	Daily dose	Agent	Daily dose
Vitamins		**Trace elements**	
Vitamin A (retinol)	0,5-1,0 mg	Chrome	50-150 µg
Vitamin B_1 (thiamine)	10-30 mg	Selenium	50-125 µg
Vitamin B_2 (riboflavin)	10-30 mg	Zinc	15-30 mg
Niacin (Vitamin B_3)	10-30 mg		
Vitamin B_6 (pyridoxin)	10-40 mg	**Minerals**	
Vitamin B_{12} (cyanocobalamin)	30-500 µg	Silicon	20-40 mg
Vitamin C (ascorbic acid)	500-1.500 mg		
Vitamin D	25-100 µg	**Quasi-vitamins**	
Natural Vitamin E	75-250 mg	Choline	40-100 mg
Of that alpha tocopherol	50-130 mg	Coenzyme Q_{10}	30-60 mg
Gama tocopherol	4-16 mg	Inositol	20-100 mg
Natural carotenoids	4-8 mg	PABA	40-60 mg
Of that alpha carotin	30-70 µg		
Beta carotin	1,0-1,9 mg	**Plant extracts**	
Cryptoxanthin	10-15 µg	Citrus flavonoids	150-300 mg
Lutein	3-9 mg	Green tea extracts	
Zeaxanthin	11-15 mg	Red wine extract	
Biotin (Vitamin H)	15-50 µg		
Folic acid (Vitamin B_9)	600-1.200 µg		
Pantothenic acid	20-80 mg		
Additional amino acids			
Phenyl alarin	500-1.500 mg		
Arginine	500-1.500 mg		
Tyrosine	400-600 mg		

Daily intake

10 – 15 ml measuring spoon morning and midday

After three months only half of the daily quantity.

10 g collagen peptides (amino acid blend) 0/1/1

Additionally

- Ferro sanol duodenal for 14 days, then 3 x week (Mo, Wed, Fri), but only when ferritin levels are < ng/ml and sTfR is > 3.0
- Jodid 50 – 100 µg, 1/0/0 only with a TSH-basal of > 2.8
- Magnesium 150 mg, 1/1/1
- 2 Tbs Norsan Omega-3 oil (1 Tbs = 1700 mg EPA/DHA), 1/1/0 after eight weeks only 1/0/0
- In the evening, 1 hour before bed 5-HTP (hydroxy tryptophan) 50 – 100 mg 0/0/1

Fig. 62: Reference ranges of customized micronutrient formulations of the HCK system for ADHD children

101 children were treated with a dietary change based on Dr. Mosetter's Glycoplan (http:www.myoreflex.de/media/downloads/Glycoplan_Ampel-Liste_Mosetter.pdf; ➡ chapter 4.7) and a customized micronutrient formulation that was composed by pharmacies based on laboratory results and evaluation via our database, and was to be taken twice a day. (fig. 62)

31 children (E) only underwent the dietary change according to Dr. Mosetter's Glycoplan, and the 30 children in the control group did not receive any additional intervention (control). A micronutrient analysis was performed on all children at the start of school, after 12 weeks, and after 24 weeks.

All children participated in sports 3-5 hours/week. Additionally, an adequate amount of health-consciousness was expected from the parents (and children), because otherwise the dietary change would not take place. For the first two weeks, the children were given 1 tsp Galactose/ribose (insulin-independent sugar) morning, midday, and evening, which very quickly reduced *food cravings, craving sweet foods*.

Chemical analyses were done on: thyroid hormone TSH-basal, parameter of 24-hour HRV-measurement (stress index, pNN50, HR, LF, as well as LF/HF-ratio), HbA1C (long-term blood sugar parameter), creatine kinase (muscle enzyme marker), intraerythrocytic lab values (HS-Omega-3 index, magnesium, zinc, selenium, Vitamins B, B2, B6, B9, holotranscobalamin), Vitamin-D, ferritin, as well as amino acids arginine, phenyl alanine, tryptophan, and tyrosine as well as the cysteine degradation product taurine.

In addition, the children were asked to complete a subjective questionnaire about ailments, inner restlessness, sleep behavior, and their ability to concentrate.

Results

With respect to the baseline data, all three patient groups were comparable (tab. 12). Just over half of the children were boys, which corresponds to the gender distribution in the literature.

Tab. 12: Baseline data of patients in all three test groups

	M & N (N = 101)	N (N = 31)	Control (N = 30)
Baseline data	Mean ± Standard deviation	Mean ± Standard deviation	Mean ± Standard deviation
Height	146,01 ± 8,401	146,03 ± 8,272	145,90 ± 7,743
Weight	39,878 ± 6,1159	40,135 ± 6,1249	39,290 ± 5,8246
Age	10,11 ± 1,476	10,26 ± 1,460	10,00 ± 1,414

Omega-3 fatty acids are part of the cell membrane and affect cell membrane permeability. They are important to brain and neural development as well as serotonin and dopamine receptor synthesis. A lack of Omega-3 fatty acids significantly increases the risk of ADHD as well as decreased cognitive performance. The literature lists an Omega-3 index of > 8% as the normal range.

At the start of school, the patients in the study presented here showed an HS-Omega-3 index in all three groups that was clearly too low (tab. 13). None of the patients showed an HS-Omega-3 index in the normal range. While a dietary change achieved only a small increase in the HS-Omega-3 index, supplementation in addition to the dietary change achieved a normal range within 12 weeks.

Tab. 13: HS-Omega-3 index in patients from all three test groups unit in %

	M & N (N = 101)	N (N = 31)	Control (N = 30)
HS-Omega-3 Index	Mean ± Standard deviation	Mean ± Standard deviation	Mean ± Standard deviation
Start of school	3,51 ± 0,67	3,24 ± 0,43	3,22 ± 0,40
After 12 weeks	8,72 ± 0,66	4,62 ± 0,55	3,16 ± 0,38

THE POWER OF MICRONUTRIENTS

The **HbA_{1c}-value** is the percentage of hemoglobin (red blood pigment) bound to glucose. It is considered the progression parameter for the blood-sugar level during the previous 8-12 weeks, and thus refers to the metabolic status. Levels below < 6% indicate no presence of diabetes mellitus, but more than 45% of the children showed an HbA_{1c}- level of > 5.7%, which is consistent with a prediabetic metabolic status. As a reminder, these children were between seven and 12 years old!

The dietary change – with and without micronutrient supplementation – clearly improved the HbA1C level in children in the upper normal range as compared to the control-group patients (tab. 14).

Tab. 14: HbA_{1c} value of patients in all three test groups in %

	M & N (N = 101)	N (N = 31)	Control (N = 30)
HbA_{1c}	Mean ± Standard deviation	Mean ± Standard deviation	Mean ± Standard deviation
Start of school	5,65 ± 0,17	5,64 ± 0,16	5,59 ± 0,18
After 12 weeks	5,32 ± 0,15	5,34 ± 0,15	5,58 ± 0,14
After 24 weeks	5,00 ± 0,15	5,07 ± 0,14	5,58 ± 0,12

In this study setting the HbA_{1c} level can also be used as a parameter for adherence to the treatment with respect to dietary change, and attests to very good adherence of children and parents (compliance).

Ferritin is the storage form of iron. Like Omega-3 fatty acids, iron is also essential to the synthesis of dopamine and other neurotransmitters. Iron is also necessary for the synthesis of thyroid hormones.

As far back as 2004, Konofal et al. were able to show that children with ADHD had lower ferritin levels than comparable healthy children. The results were verified via a meta-analysis by Tseng et al. in 2018. Ferritin levels of < 50 ng/ml show a deficiency in functional iron. The results of the study presented here also show a ferritin level at the start of school in all three groups that is significantly too low (tab. 15).

Tab. 15: Ferritin concentration in patients from all three test groups in ng/ml

	M & N (N = 101)	N (N = 31)	Control (N = 30)
Ferritin	Mean ± Standard deviation	Mean ± Standard deviation	Mean ± Standard deviation
Start of school	36,56 ± 4,76	33,11 ± 4,96	36,61 ± 3,81
After 12 weeks	64,64 ± 9,40	37,13 ± 3,26	35,51 ± 3,94
After 24 weeks	78,58 ± 5,54	39,89 ± 2,78	34,59 ± 4,00

The study presented here was able to show that a customized ferritin supplement could not only significantly improve the lab values of children with ADHD, but also their symptoms, while a dietary change alone resulted in only a marginal improvement.

Another trace element the literature lists as being linked to ADHD is zinc. There is no consistent data on the significance of magnesium in ADHD. To date no connection has been shown to copper, selenium, manganese, and iodine. We will therefore not go into further detail about trace elements.

A connection between Vitamin D deficiency and poor neurocognition is mentioned; **Vitamin D** regulates neural development, monoamine and serotonin concentrations, which in turn are relevant to ADHD.

The literature consistently describes low Vitamin D levels in children with ADHD (Kotsis, 2019), which coincides with the results from this study: At the start of school the children in all three groups showed Vitamin D levels that were much too low (24.88 ± 7.3 nmol/l; the normal level is > 75 nmol/l; Holick, 2007; tab. 16). Standard Vitamin D levels could only be achieved through supplementation.

Tab. 16: Vitamin D concentration in patients from all three test groups in nmol/l

	M & N (N = 101)	N (N = 31)	Control (N = 30)
Vitamin D	Mean ± Standard deviation	Mean ± Standard deviation	Mean ± Standard deviation
Start of school	26,72 ± 5,69	24,32 ± 5,50	24,33 ± 4,63
After 24 weeks	81,87 ± 6,46	22,81 ± 5,73	22,80 ± 4,43

The **thyroid-stimulating hormone** (TSH) is an important parameter for thyroid function, which in turn affects vegetative symptoms as well as the metabolism. Neural maturation in children as well as their skeletal and body growth are regulated by thyroid hormones. Synthesis of thyroid hormones requires selenium, iodine, zinc, Vitamin D, and Omega-3 fatty acids, etc.

A deficient TSH basal value (< 1.3 μIU/ml, with resulting hyperthyroidism) correlates with sustained tension. An elevated TSH-basal value (> 250 μIU/ml; with resulting hypothyroidism) is associated with "sleeping-pill mentality" and difficulty concentrating. A sense of wellbeing exists at a TSH-basal value between 1.6 and 2.2 μIU/ml. Dr. Wienecke was able to show this (2018), but it is contrary to information from the area of conventional medicine, whose definition of the "normal range" is much more generous.

At the start of school, only two of the three groups showed mean values outside the normal range (tab. 17). However, TSH-basal values in the upper normal range can be lowered to optimal well-being-values via supplementation with micronutrients that are essential to hormone production, while no change in thyroid function can be achieved without supplementation with trace elements.

Tab 17: TSH-basal values in patients from all three test groups in μIU/ml

	M & N (N = 101)	N (N = 31)	Control (N = 30)
TSH-basal	Mean ± Standard deviation	Mean ± Standard deviation	Mean ± Standard deviation
Start of school	2,30 ± 1,10	2,56 ± 1,32	2,68 ± 1,22
After 12 weeks	2,18 ± 0,44	2,55 ± 1,27	2,71 ± 1,19
After 24 weeks	2,11 ± 0,23	2,55 ± 1,31	2,77 ± 1,22

At the start of school, all children showed a **stress index** that was too high. It was lowered with a dietary change and even more successfully with a dietary change plus micronutrient supplementation (tab. 18).

Tab. 18: Stress index in patients from all three test groups

	M & N (N = 101)	N (N = 31)	Control (N = 30)
Stress index	Mean ± Standard deviation	Mean ± Standard deviation	Mean ± Standard deviation
Start of school	378,58 ± 56,15	368,94 ± 54,47	393,37 ± 40,72
After 12 weeks	215,23 ± 43,36	307,29 ± 54,26	345,97 ± 29,89
After 24 weeks	180,13 ± 30,20	287,23 ± 47,87	322,10 ± 31,11

Fig. 63 clearly shows that the stress index dropped significantly with normalization of TSH basal values in patients from the micronutrient-diet group, which emphasizes the importance of properly adjusted Type equation here.thyroid hormone levels.

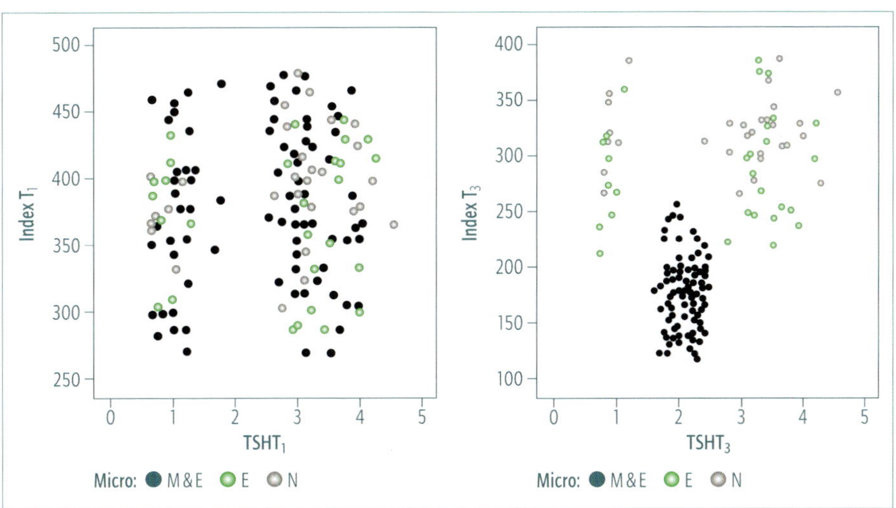

Fig. 63: Stress index and TSH basal value at the time of measurement at the start of school (left) and after 24 weeks (right) in all three test groups

While no child showed a pNN50 value in the target range at the start of school, it was accomplished with a dietary change plus micronutrient supplementation. Activity of the parasympathetic nervous system is now in the desirable range and the ADHD patients' symptoms are significantly improved (tab. 19).

Tab. 19: pNN50 value of patients from all three test groups

	M & N (N = 101)	N (N = 31)	Control (N = 30)
pNN50	Mean ± Standard deviation	Mean ± Standard deviation	Mean ± Standard deviation
Start of school	7,49 ± 0,67	7,66 ± 0,57	6,83 ± 0,55
After 12 weeks	13,81 ± 0,71	7,34 ± 1,00	7,25 ± 0,76
After 24 weeks	15,24 ± 0,70	7,95 ± 0,75	6,79 ± 0,52

Fig. 64 clearly shows that the pNN50 value significantly increased with normalization of the TSH basal value in patients from the micronutrient diet group, which emphasizes the vegetative nervous system's dependency on a properly adjusted thyroid hormone level.

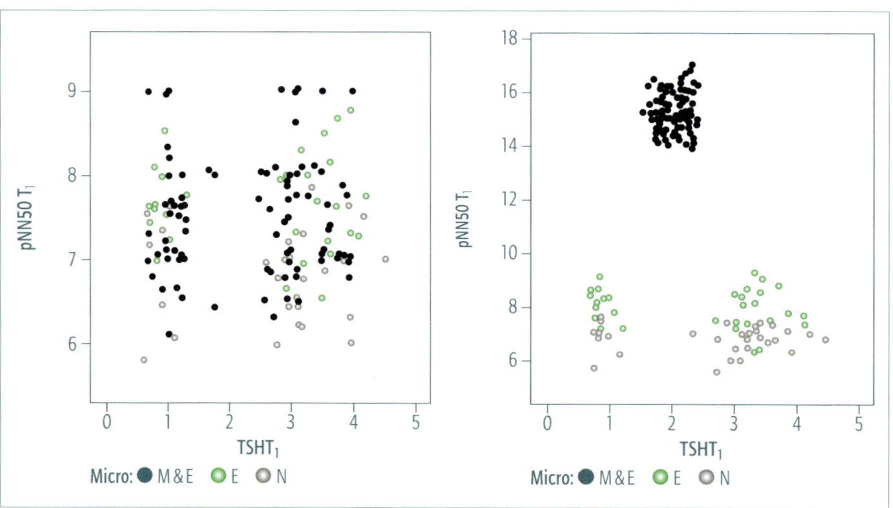

Fig. 64: pNN50 and TSH basal value at the time of measurement before the start of school (left), and after 24 weeks (right) in all three test groups.

The LF/HF-ratio was reduced in all three groups and thus shows a shift towards the parasympathetic nervous system, whereby only the values from the micronutrient diet group are impressive and reflect a definite reduction in the activity of the parasympathetic nervous system (tab. 20).

Tab. 20: Vegetative quotient (LF/HF-ratio) of patients in all three test groups

	M & N (N = 101)	N (N = 31)	Control (N = 30)
LF/HF-Ratio	Mean ± Standard deviation	Mean ± Standard deviation	Mean ± Standard deviation
Start of school	7,32 ± 0,18	7,24 ± 1,17	7,25 ± 0,91
After 12 weeks	1,82 ± 0,17	6,64 ± 0,07	6,90 ± 0,80
After 24 weeks	1,60 ± 0,13	6,09 ± 1,00	6,18 ± 0,69

Fig. 65 clearly shows that the LF/HF ratio in patients from the micronutrient diet group significantly dropped with normalization of TSH basal values, which emphasizes the vegetative nervous system's dependency on properly adjusted thyroid hormone values.

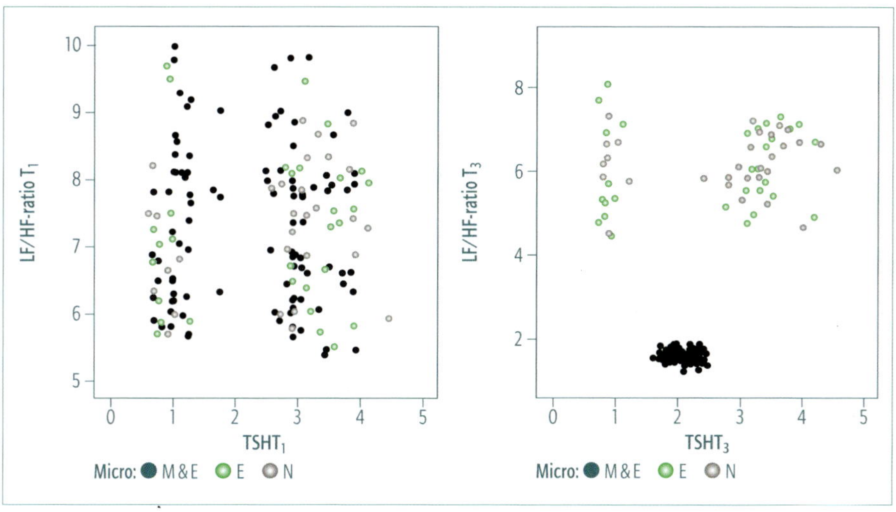

Fig. 65: LF/HF ratio and TSH basal value at the time of measurement before the start of school (left), and after 24 weeks (right) in all three test groups.

Overall, the parameters of the 24-hour HRV measurement show stabilization of the vegetative nervous system that correlates with the subjective statements regarding symptoms.

That micronutrient therapy and dietary change is not just "lab parameter cosmetics" is supported, on the one hand, by the results of the 24-hour HRV measurement, and on the other hand, also by the findings from subjective patient assessment.

Of note is that the 101 children in the combination-therapy group describe their own physical/mental state as significantly more positive than the children in the other test groups. Patient satisfaction is an important factor not only in therapy adherence but also in participation and quality of life.

CONCLUSION

This study showed that a deficiency-oriented micronutrient therapy in children with ADHD combined with a dietary change resulted in a significant improvement of symptoms, subjective sense of wellbeing, and thus participation in life. The results from this evidence-based, retrospective study show new, very successful paths in micronutrient therapy for children with ADHD.

6.3.1.2 Case study of a 5-year-old boy

The anamnesis reveals: a five-year-old boy, height 112 cm, weight 19.6 kg, within the age-appropriate range. During the interview he, or rather his parents, state that he suffers from increased fatigue, lack of motivation, inner restlessness, extreme difficulty concentrating, tendency towards hypothyroidism, mood fluctuations, night sweats, poor sleep.

He likes to eat sweets, has pollen-/dust allergies, plays club soccer 2-3 x a week. He is currently given homeopathic products to improve his physical/mental state. His pediatrician recommends methylphenidate (Ritalin).

The individual micronutrient- or energy requirement was ascertained according to the previously described testing modules and included:

- CBC without differential, plus TSH basal value, ferritin, TfR, liver and kidney values;
- Functional energy metabolism (first urine of the morning);
- Intracellular micronutrient concentrations as well as Omega-3-index;
- Amino acid determination;
- 48-hour HRVmeasurement.

The results we compiled from blood and urine analyses are shown in concise form without additional detailed remarks (fig. 66)

5-year-old boy: ADHD diagnosis before the start of treatment

TSH basal value	3,15 µlU/ml	Transferrin saturation	19 %
fT_3: 4,61 pg/ml, fT_4: 1,23 pg/ml		Transferrin receptor soluble	1,83 mg/l
Ferritin	19,9 ng/ml	Urea	31 mg/dl
Uric acid	4,1 mg/d	Sodium	137 nmol/l
HbA_{1c}	5,6 % (33,1 mM/M)	Calcium	2,14 nmol/l
CRP	< 0,30 mg/dl	Hematocrit	46,4 %
GOT	48 U/l	Leucocytes	5,20 Tsd/µl
GPT	24 U/l	Erythrocytes	5,11 Mio/µl
GGT	19 U/l	Hemoglobin	14,8 g/dl
CK	112 U/l	MCV	96,2 fl
Alkaline phosphatase	56 U/l	MCU	31,5 pg
Bilirubin total	0,38 mg/dl	MCH	34,1 g/dl
Iron	169 µg/dl	Thrombocytes	185 Tsd/µl
Transferrin	280 mg/dl		

Fig 66: Lab values of a five-year-old with ADHD before the start of treatment

Iron absorption by the body's cells is controlled via expression of transferrin receptors (TfR). When there is an iron deficiency the sTfR concentration increases even before the hemoglobin concentration significantly decreases. The functional iron status can therefore be tested via the sTfR concentration while ferritin offers information about the status of iron stores.

The low TSH basal value corresponds to the clinical data of increased fatigue, lack of motivation, and a tendency to hypothyroidism. Still of note are the decreased intracellular levels of magnesium and zinc, a significantly decreased HS-Omega-3 index, as well as – shocking in a five-year-old – an active insulin resistance.

The child's nutritional behavior is based on the basic principles of Dr. Mosetter's Glycoplan (➡ chapter 4.7). In addition, the boy is taking the following micronutrient formulation (fig. 67).

5-year-old boy: ADHD diagnosis – micronutrient formulation

Agent	Daily dose	Agent	Daily dose
Vitamins		**Trace elements**	
Vitamin A (retinol)	1 mg	Chrome	150 µg
Vitamin B_1 (thiamine)	15 mg	Selenium	125 µg
Vitamin B_2 (riboflavin)	15 mg	Zinc	24 mg
Niacin (Vitamin B_3)	40 mg		
Vitamin B_6 (pyridoxin)	40 mg	**Minerals**	
Vitamin B_{12} (cyanocobalamin)	575 µg	Calcium	200 mg
Vitamin C (ascorbic acid)	1.250 mg	Silicon	20 mg
Vitamin D	120 µg		
Natural Vitamin E	100 mg	**Quasi-vitamins**	
Of that alpha tocopherol	107,5 mg	Choline	100 mg
Gama tocopherol	42,5 mg	Coenzyme Q_{10}	30 mg
Natural carotenoids	8 mg	Inositol	70 mg
Of that alpha carotene	70 µg	PABA	30 mg
Beta carotene	1,9 mg		
Cryptoxanthin	15 µg	**Plant extracts**	
Lutein	4 mg	Citrus flavonoids	245,9 mg
Zeaxanthin	11 mg		
Biotin (Vitamin H)	100 µg		
Folic acid (Vitamin B_9)	1.200 µg		
Pantothenic acid	30 mg		

Daily intake
Micronutrient blend
1 measuring scoop morning and midday, after 3 months 1 measuring scoop only in the morning
Additionally:
Iron, e.g. ferro sanol duodenal, 14 days in the evening, then 3 x a week (Mon, Wed, Fri)
Iodine: e.g. Jodid 100, 1/0/0 dosage
Magnesium (e.g. 30-40 mg of a magnesium aspartate), 2/2/4 dosage
Omega-3 oil (1 Tbs = 1700 mg EPA/DHA), 1/0/0 dosage
5-HTP (100 mg) 1 hour before bed, 0/0/1 dosage
Galactose as treatment (1 tsp) 1/1/1/ dosage

Fig. 67: Micronutrient formulation for a five-year-old with ADHD before the start of therapy

Fig. 68, 69, and 70 show the energy, micronutrient, and amino acid profiles before and after the start of treatment.

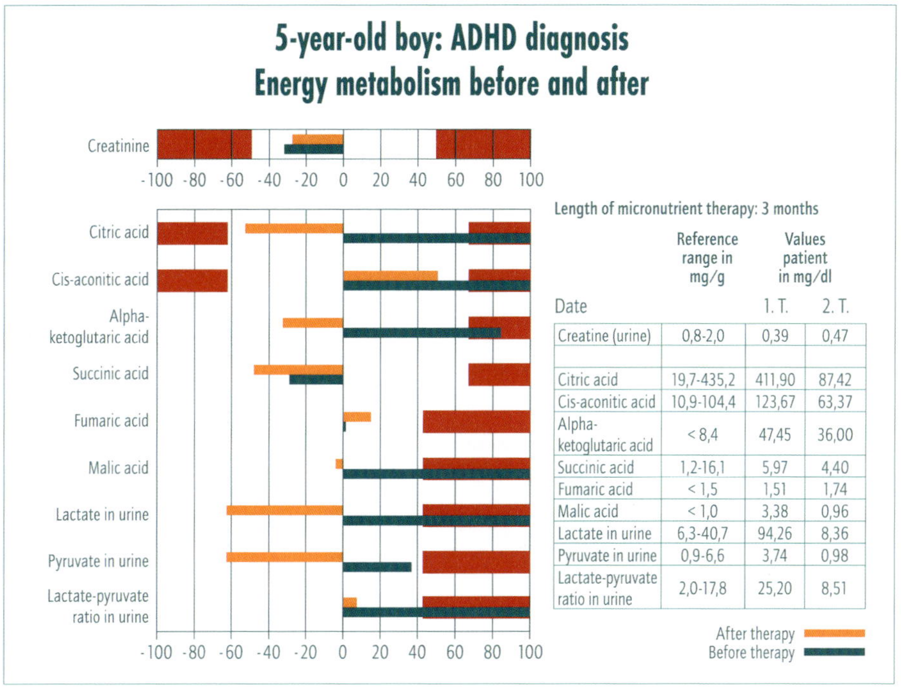

Date	Reference range in mg/g	Values patient in mg/dl 1. T.	2. T.
Creatine (urine)	0,8-2,0	0,39	0,47
Citric acid	19,7-435,2	411,90	87,42
Cis-aconitic acid	10,9-104,4	123,67	63,37
Alpha-ketoglutaric acid	< 8,4	47,45	36,00
Succinic acid	1,2-16,1	5,97	4,40
Fumaric acid	< 1,5	1,51	1,74
Malic acid	< 1,0	3,38	0,96
Lactate in urine	6,3-40,7	94,26	8,36
Pyruvate in urine	0,9-6,6	3,74	0,98
Lactate-pyruvate ratio in urine	2,0-17,8	25,20	8,51

Fig. 68: Energy profile of a five-year-old with ADHD before and after therapy

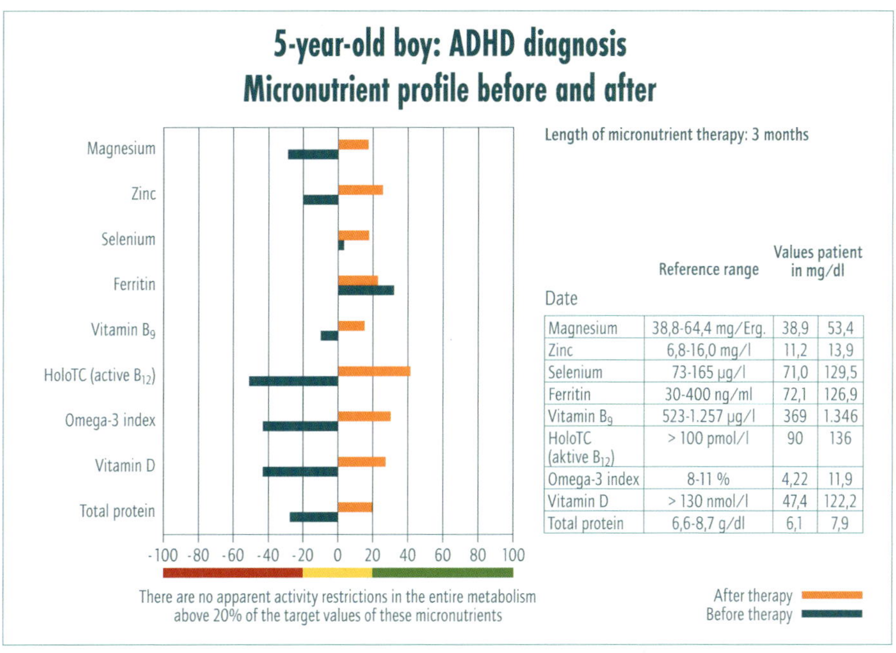

Date	Reference range	Values patient in mg/dl	
Magnesium	38,8-64,4 mg/Erg.	38,9	53,4
Zinc	6,8-16,0 mg/l	11,2	13,9
Selenium	73-165 µg/l	71,0	129,5
Ferritin	30-400 ng/ml	72,1	126,9
Vitamin B_9	523-1.257 µg/l	369	1.346
HoloTC (aktive B_{12})	> 100 pmol/l	90	136
Omega-3 index	8-11 %	4,22	11,9
Vitamin D	> 130 nmol/l	47,4	122,2
Total protein	6,6-8,7 g/dl	6,1	7,9

Fig 69: Micronutrient profile of a five-year-old with ADHD before and after therapy

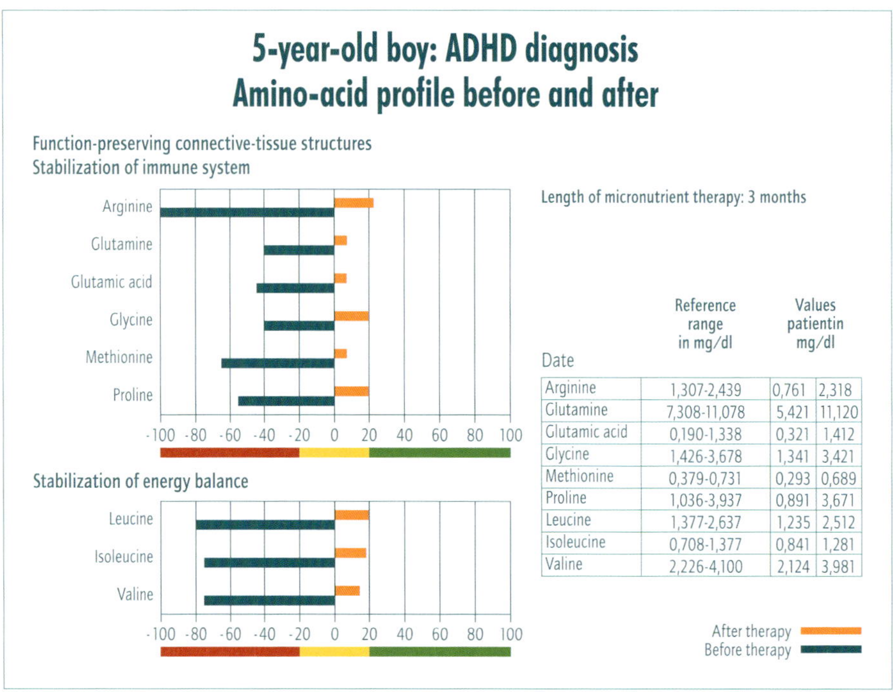

Date	Reference range in mg/dl	Values patientin mg/dl	
Arginine	1,307-2,439	0,761	2,318
Glutamine	7,308-11,078	5,421	11,120
Glutamic acid	0,190-1,338	0,321	1,412
Glycine	1,426-3,678	1,341	3,421
Methionine	0,379-0,731	0,293	0,689
Proline	1,036-3,937	0,891	3,671
Leucine	1,377-2,637	1,235	2,512
Isoleucine	0,708-1,377	0,841	1,281
Valine	2,226-4,100	2,124	3,981

Fig. 70: Amino acid profile of a five-year-old with ADHD before and after therapy

The measured results also reflect the mental and physical state: After three months, the increased fatigue, lack of motivation, inner restlessness, extreme difficulty concentrating, severe mood swings, night sweats, and poor sleep changed to good mental and physical performance ability, a very balanced mood, no night sweats, and a positive surprise for the parents with respect to the change in their son's behavior.

The results from the 48-hour HRV analysis show a significant reduction in the stress index (fig. 71) and a significant rise in the pNN50 (fig. 72). The vegetative quotient (fig. 73) has come down significantly and shows the extraordinary impact of a customized micronutrient intake on the balance of the vegetative nervous system.

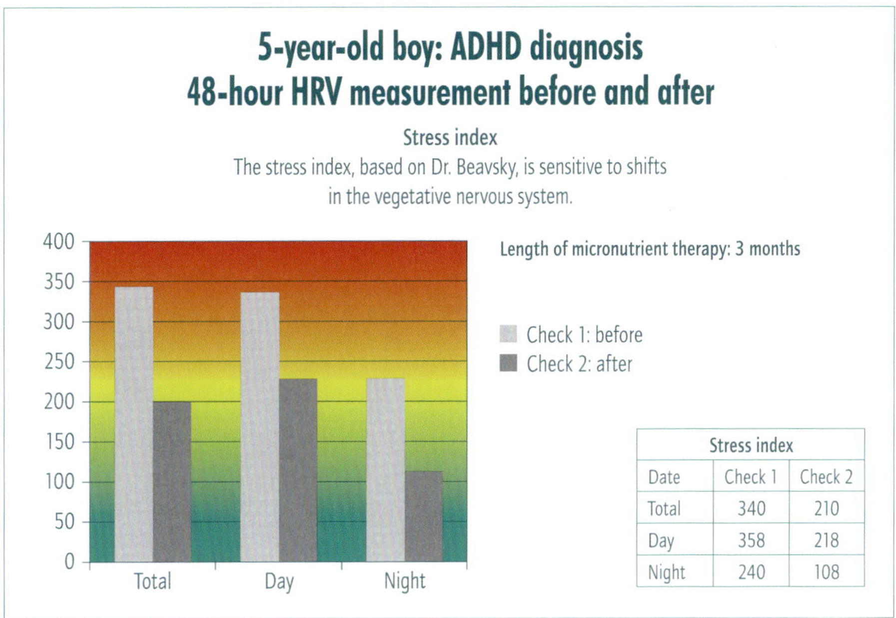

Stress index		
Date	Check 1	Check 2
Total	340	210
Day	358	218
Night	240	108

Fig. 71: Stress index in a five-year-old with ADHD before and after therapy

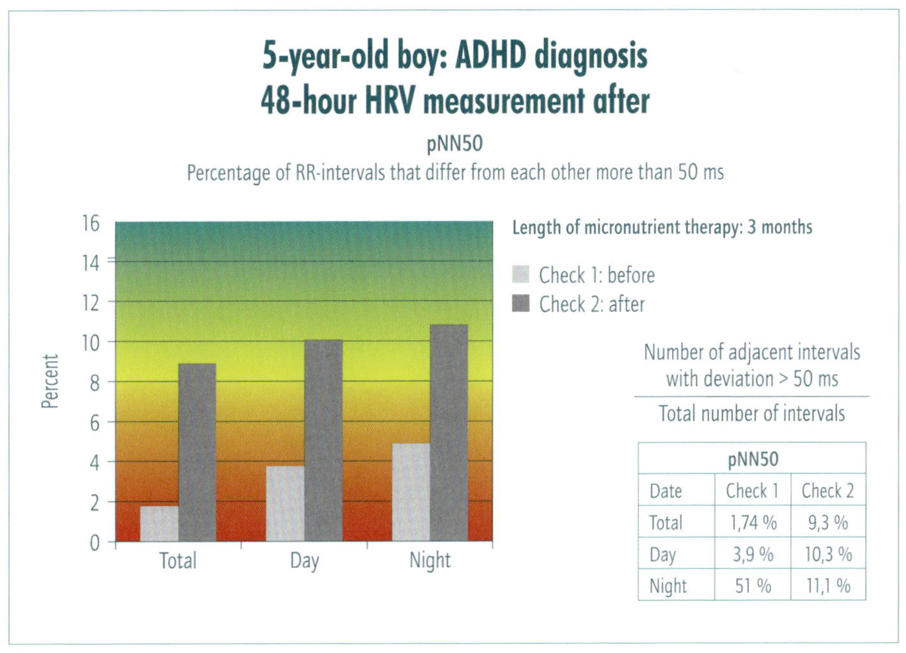

pNN50		
Date	Check 1	Check 2
Total	1,74 %	9,3 %
Day	3,9 %	10,3 %
Night	51 %	11,1 %

Fig. 72: pNN50 of a five-year-old with ADHD before and after therapy

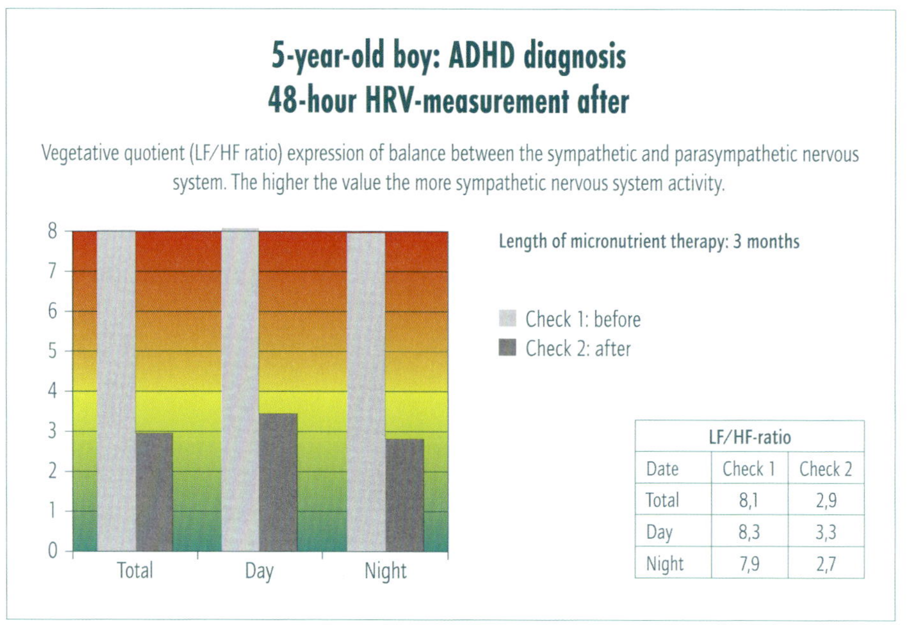

LF/HF-ratio		
Date	Check 1	Check 2
Total	8,1	2,9
Day	8,3	3,3
Night	7,9	2,7

Fig. 73: Vegetative quotient of a five-year-old with ADHD before and after therapy

6.3.1.3 Micronutrients and ADHD

The following graphic illustrations show how a targeted, customized micronutrient supplementation (fig. 74, 75, 76, 77) can have a positive effect on the physical and mental state of children with ADHD.

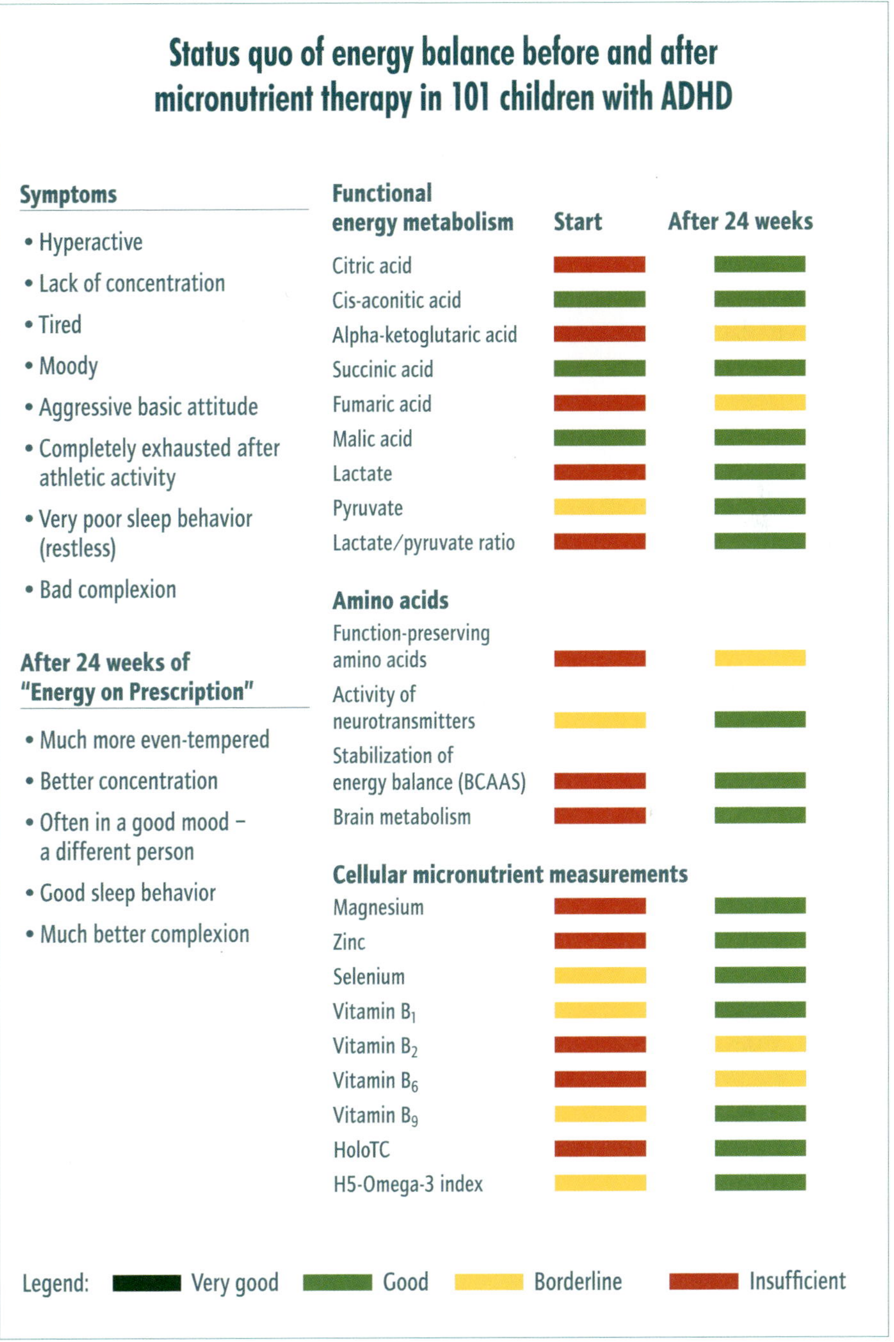

Fig. 74: Status quo of energy balance before and after micronutrient therapy in 101 children with ADHD

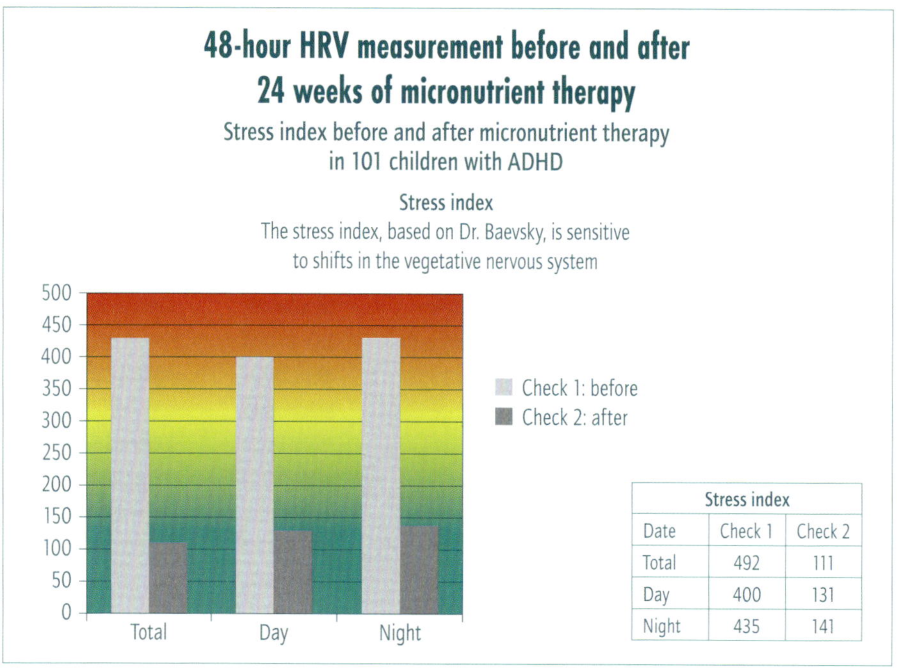

Stress index		
Date	Check 1	Check 2
Total	492	111
Day	400	131
Night	435	141

Fig. 75: Stress index before and after micronutrient therapy in 101 children with ADHD

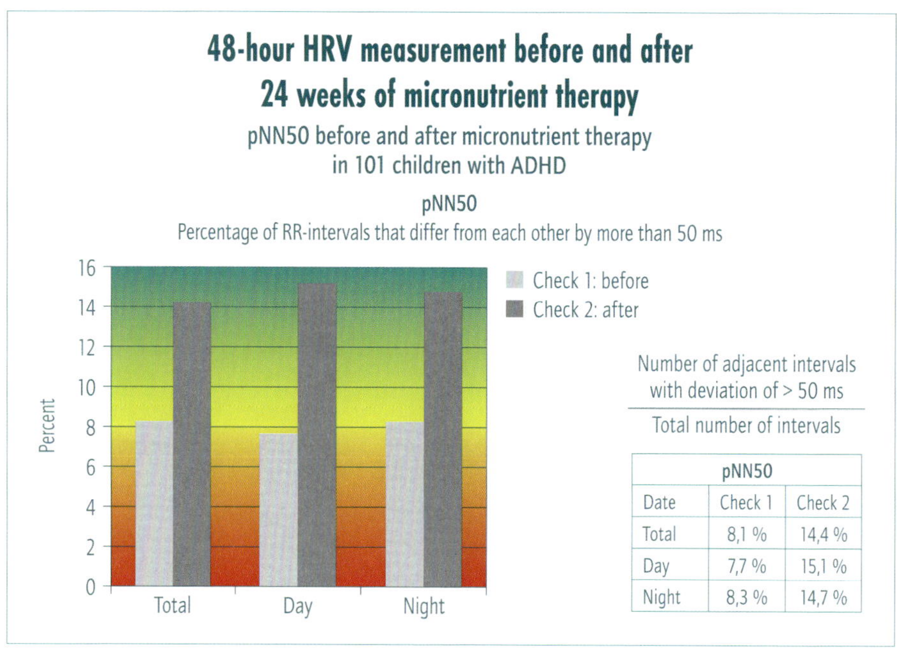

pNN50		
Date	Check 1	Check 2
Total	8,1 %	14,4 %
Day	7,7 %	15,1 %
Night	8,3 %	14,7 %

Fig. 76: pNN50 before and after micronutrient therapy in 101 children with ADHD

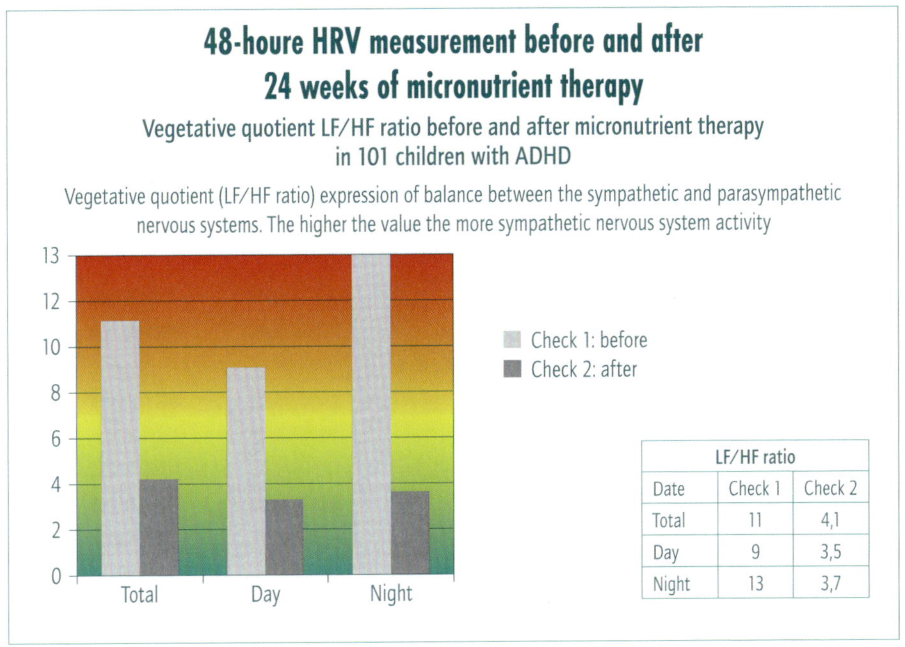

LF/HF ratio		
Date	Check 1	Check 2
Total	11	4,1
Day	9	3,5
Night	13	3,7

Fig. 77: Vegetative quotient before and after micronutrient therapy in 101 children with ADHD

6.3.2 Chronic pain

Pain from various causes is a widespread health problem that causes many people to abuse over-the-counter pain medicines. And yet pain caused by different illnesses (rheumatism, fibromyalgia) can be reduced without side effects with a customized micronutrient formulation (fig. 78, 79, 80 and 81).

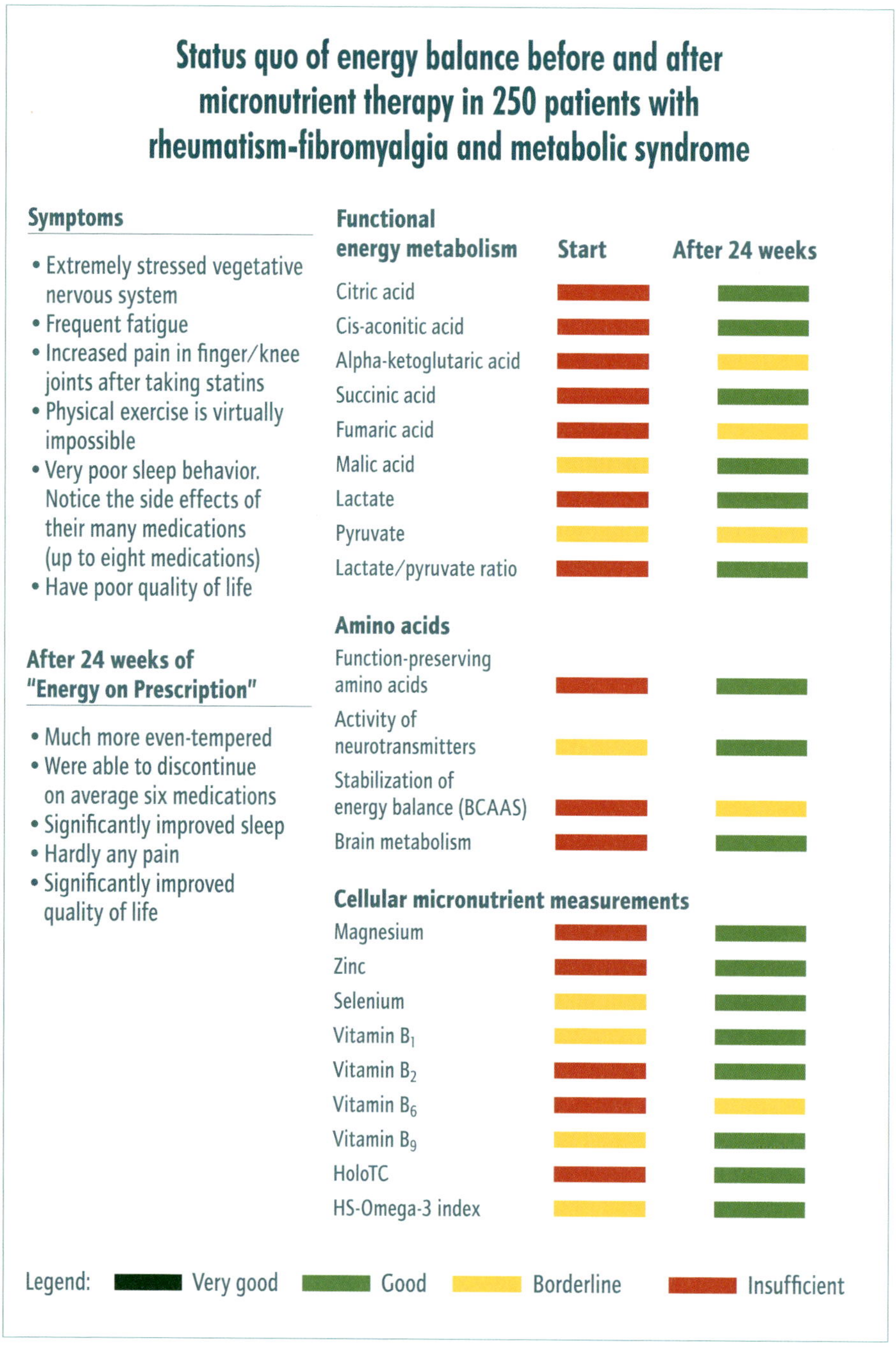

Fig. 78: Status quo of energy balance before and after micronutrient therapy in 250 patients with rheumatism, fibromyalgia, and metabolic syndrome

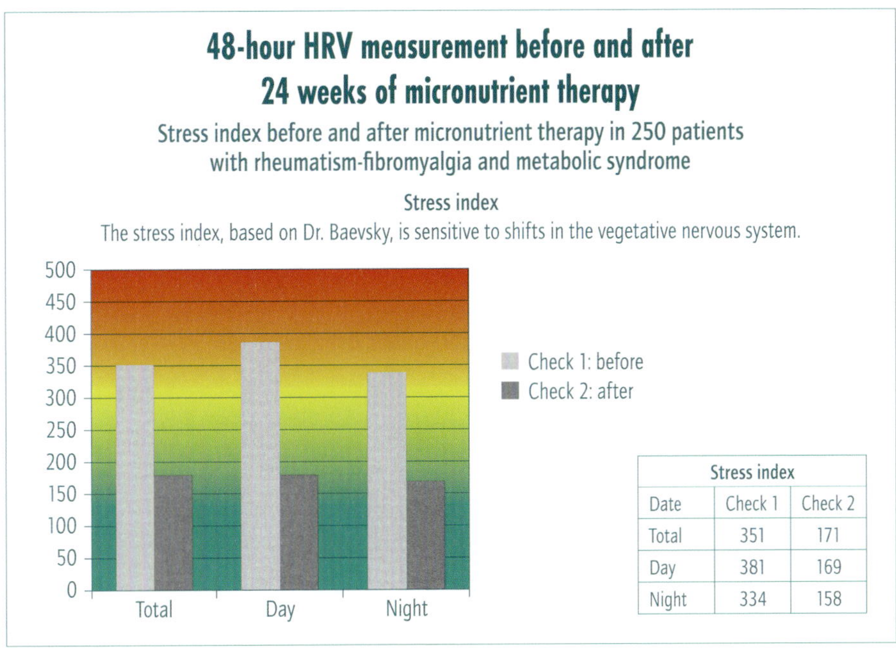

Stress index		
Date	Check 1	Check 2
Total	351	171
Day	381	169
Night	334	158

Fig. 79: Stress index before and after micronutrient therapy in 250 patients with rheumatism, fibromyalgia, and metabolic syndrome

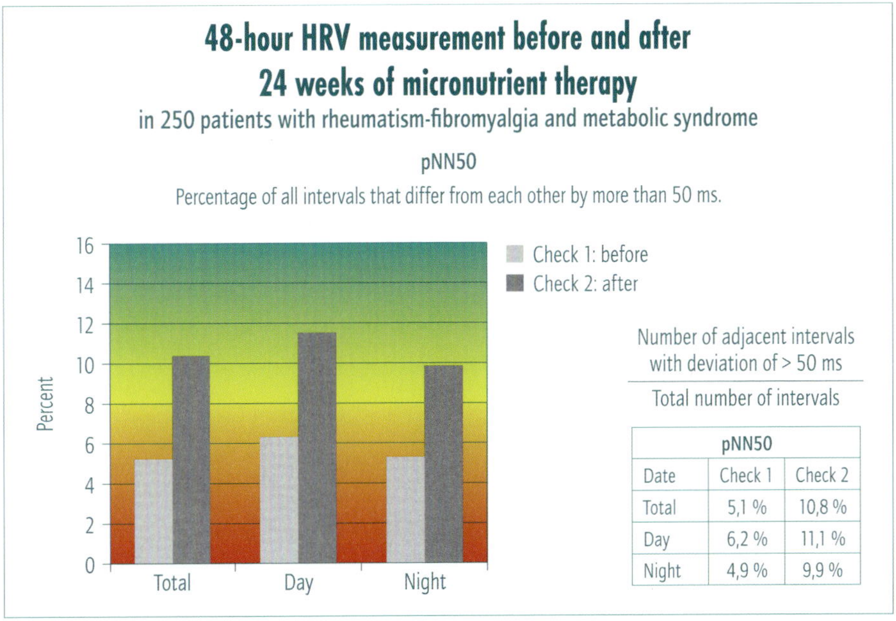

pNN50		
Date	Check 1	Check 2
Total	5,1 %	10,8 %
Day	6,2 %	11,1 %
Night	4,9 %	9,9 %

Fig. 80: pNN50 before and after micronutrient therapy in 250 patients with rheumatism, fibromyalgia, and metabolic syndrome

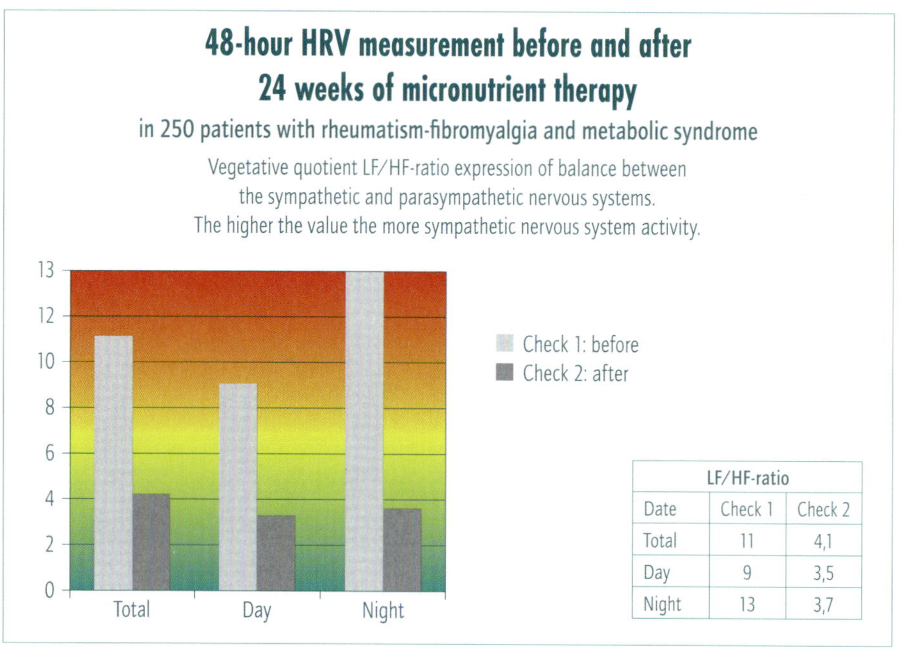

LF/HF-ratio		
Date	Check 1	Check 2
Total	11	4,1
Day	9	3,5
Night	13	3,7

Fig. 81: Vegetative quotient before and after micronutrient therapy in 250 patients with rheumatism, fibromyalgia, and metabolic syndrome

A total of 1,250 patients with similar anamnesis profiles were tested in recent years. The shown analysis included only 250 patients as the others were only entered into the database after the diagrams had been created. All data are documented in the database along with the respective reference or target values of the individual micronutrient concentrations. The goal is to correct the biochemical disturbances via a targeted micronutrient formulation.

6.3.2.1 Case study of a 76-year-old male

The patient suffers from hypertension, type-1 diabetes, rheumatism, hypercholesterolemia, and is slightly overweight. His height is 181 cm, weight is 95 kg, his BMI of 29 is at the upper end of the age-appropriate normal range. He is also suspected of suffering from hypothyroidism and reports trouble sleeping as well as very little physical activity. Even just taking a walk is too much.

The clinical exam showed severe degeneration of the active and passive locomotor system with significantly impaired mobility in finger joints and wrists. His medications prior to micronutrient therapy consisted of:

- Pleon 5 mg (active ingredient sulfasalazine): 1-1-1;
- Prednisolone acis 5 mg: 1-0-0-5;
- Metformin Lich 850 mg: 1-0-1;
- Simvastatin 40 mg: 1-0-0;
- Metoprolol 50: 1-0-0;
- Indomethacin 50 mg: 0-0-1;
- Omeprazole: 1-0-0.

Previous medical exams result in the recommendation of two additional pain relievers. But the patient absolutely rejects that idea. The special blood and urine tests and the comparison with cluster patients in his age group show massive deficiencies in his micronutrient balance. The functional energy metabolism shows significantly limited activity of certain enzymes. Fig. 82 shows the results from the blood and urine tests prior to the start of micronutrient therapy.

The blood results show no abnormalities in the borrelia serology. The Western blot test (checking for possible tickborne disease) and EBV serology (checking for possible presence of mononucleosis) are negative.

Lab values of a 76-year-old patient before micronutrient therapy

TSH-basal value (Thyroid antibodies TPO, TRAb normal)	2,91 µIU/ml
Ferritin	131 ng/ml
Leuco	8,90 Tsd/µl
RBC	5,0 Mio/µl
Hb	14,8 g/dl
Uric acid	6,7 mg/d
Chol	179 mg/dl, with medication
HDL	79 mg/dl
LDL	92 mg/dl

HbA_{1c}	5,9 % (40,9 mM/M) with medication
CRP	1,9 mg/dlhemoglobin
Liver and kidney values	normal
Auto-antibodies against thyroid peroxidase (Microsomal AB) (reference range 34)	14.4 U/ml
TSH receptor (Reference range 0.92 negative)	< 0,4 U/l
Borrelia serology	Negative
Western blot test	Negative
EBV-serology	Negative

Optimal ferritin levels at 131 ng/ml, additionally measured sTfR (transferrin receptor; 1.6 no functional iron deficiency)
Intracellular analyses Mg, Zn, Se, B_9 and HoloTC (active Vitamin B_{12}) show dramatically low concentrations, particularly notable are the extremely low intracellular:

- Mg-concentration at 35.7 mg/l ery. (optimal: 55 mg/l ery.)
- Zn-concentration at 10.27 mg/l ery. (optimal: 13.4 mg/l ery)

Significantly low HS-Omega-3 index: 3.9% FS
HbA_{1C}-value: 5.9% (40.9 mM/M) due to medication
Hypercholesterolemia normalized with medication
Severely limited activity of certain enzymes in functional energy metabolism

Fig. 82: Lab parameters for a 76-year-old patient before micronutrient therapy

Results for individual amino acids show urgent need for optimization in all areas. This is particularly the case for amino acids that negatively affect the entire spectrum of the myofascial system. The significantly lowered L-tryptophan concentration at 0.89 mg/dl is very striking. Healthy sleep behavior is only possible at a tryptophan concentration of 1.7 m/dl. Since the 76-year-old man has chronic inflammation, he cannot receive L-tryptophan as this can cause toxic reactions (kynurenine accumulation).

He is therefore given the amino acid 5-HTP, which is biochemically direct-acting by bypassing the blood-cerebral barrier and does not trigger a negative reaction.

The patient receives a customized micronutrient formulation involving the HCK modular system: daily dose 35 ml, replenishment phase 14 ml in the morning and 14 ml midday; after an eight-week transition to the maintenance phase with 20 ml in the morning (fig. 83).

Micronutrient therapy of a 76-year-old patient

Agent	Daily dose	Agent	Daily dose
Vitamins		**Trace elements**	
Vitamin A (retinol)	1 mg	Chrome	400 µg
Vitamin B_1 (thiamine)	30 mg	Selenium	230 µg
Vitamin B_2 (riboflavin)	30 mg	Zinc	50 mg
Vitamin B_6 (pyridoxin)	60 mg		
Vitamin B_{12} (cyanocobalamin)	2.100 µg	**Minerals**	
Vitamin C (ascorbic acid)	3.000 mg	Silicon	40 mg
Vitamin D	200 µg		
Natural Vitamin E	150,4 mg	**Quasi vitamins**	
Of that alpha tocopherol	86,5 mg	Choline	240 mg
Gama tocopherol	35,5 mg	Coenzyme Q_{10}	120 mg
Natural carotenoids	8 mg	Inositol	180 mg
Of that alpha carotin	70 µg	PABA	60 mg
Beta-Carotin	1,9 mg		
Cryptoxanthin	15 µg	**Plant extracts**	
Lutein	6 mg	Citrus flavonoids	275,9 mg
Zeaxanthin	15 mg		
Biotin (Vitamin H)	150 µg	**Additionally**	
Folic acid (Vitamin B_9)	1.600 µg	Jodid	100 µg, 1/0/0
Niacin (Vitamin B_3)	30 mg	Magnesium	300 mg, 0/0/1
Pantothenic acid	60 mg	Norsan Omega-3 oil (1 Tbs = 1.700 mg EPA/DHA), 1/1/0	
Additional amino acids in special cases		5-HTP (hydroxytryptophan) Griffonia	100 mg, 1/1/1
Arginine	2.000 mg		
MSM	2.000 mg		

Fig. 83: Micronutrient therapy of a 76-year-old patient

The following image results after 12 weeks of targeted micronutrient therapy: The targeted intake of the listed micronutrient intake has balanced the deficiencies and the organism is once again able to regulate itself. The targeted ingestion of the amino acid 5-HTP with 100 mg each in the morning, midday, and evening has resulted in significant pain reduction within three weeks. The results of the functional energy metabolism analysis document the effectiveness of the ingested micronutrient formulation (fig. 84).

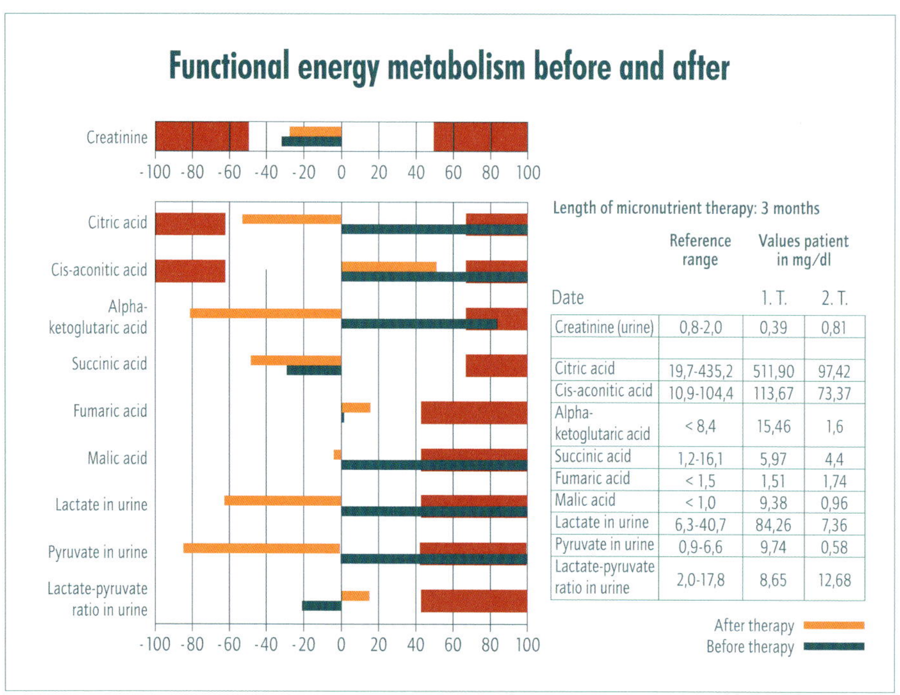

Date	Reference range	Values patient in mg/dl 1. T.	2. T.
Creatinine (urine)	0,8-2,0	0,39	0,81
Citric acid	19,7-435,2	511,90	97,42
Cis-aconitic acid	10,9-104,4	113,67	73,37
Alpha-ketoglutaric acid	< 8,4	15,46	1,6
Succinic acid	1,2-16,1	5,97	4,4
Fumaric acid	< 1,5	1,51	1,74
Malic acid	< 1,0	9,38	0,96
Lactate in urine	6,3-40,7	84,26	7,36
Pyruvate in urine	0,9-6,6	9,74	0,58
Lactate-pyruvate ratio in urine	2,0-17,8	8,65	12,68

Fig. 84: Status quo of the energy balance before and after micronutrient therapy of the 76-year-old patient

The acute phase protein (CRP) was verifiably normalized. The pain relievers could be discontinued, the significantly elevated cholesterol levels could be lowered from 310 mg/dl fasting to 185 mg/dl specifically via the targeted intake of the high dose of Omega-3 fatty acids. The considerable side-effects from the medications taken with respect to increasing muscle pain have been eliminated. The TSH-basal value was reduced and with 1.9 µIU/ml (➨ chapter 3.1) was now in the "wellbeing" reference range.

Stress index, pNN50, as well as the vegetative quotient (fig. 85, 86, and 87) that were analyzed with the 24-hour HRV measurement have normalized and shown a much better balance of the vegetative nervous system.

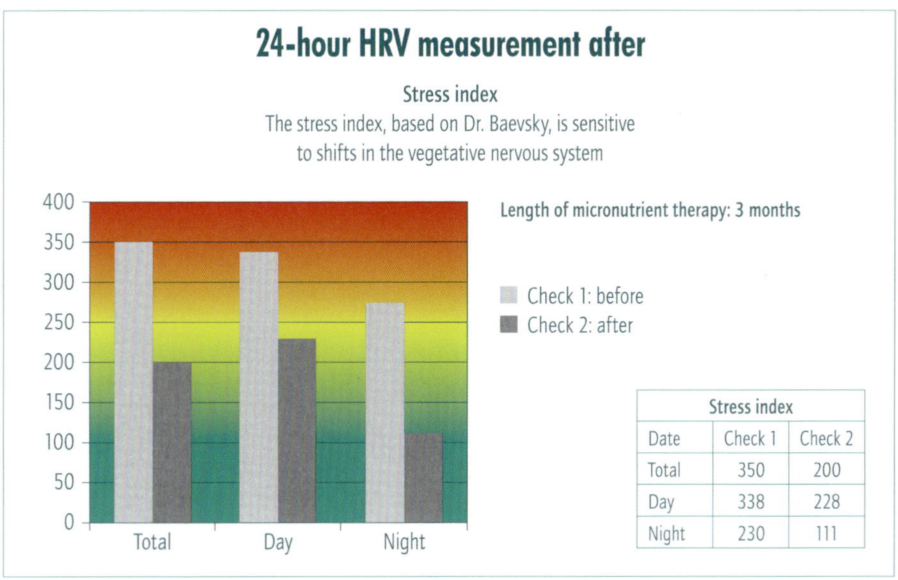

Stress index		
Date	Check 1	Check 2
Total	350	200
Day	338	228
Night	230	111

Fig. 85: Stress index before and after micronutrient therapy in a 76-year-old patient

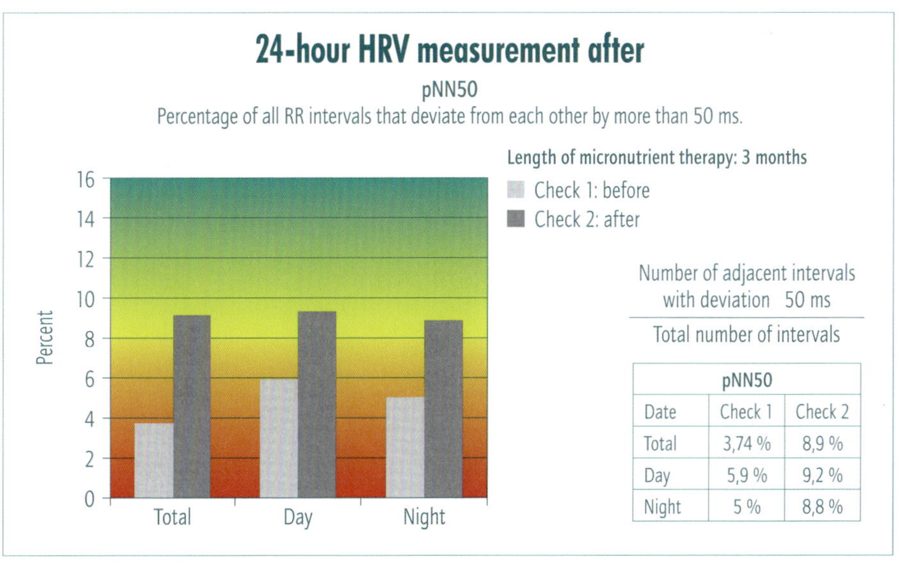

pNN50		
Date	Check 1	Check 2
Total	3,74 %	8,9 %
Day	5,9 %	9,2 %
Night	5 %	8,8 %

Fig. 86: pNN50 before and after the micronutrient therapy in a 76-year-old patient

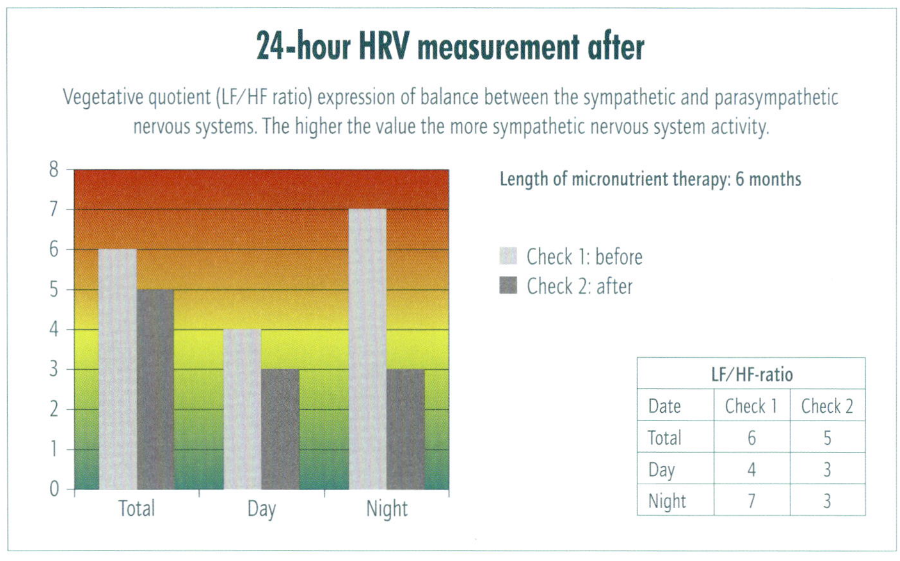

LF/HF-ratio		
Date	Check 1	Check 2
Total	6	5
Day	4	3
Night	7	3

Fig. 87: Vegetative quotient before and after micronutrient therapy in the 76-year-old male patient

The patient is also doing much better subjectively. He is rarely tired, no longer has night sweats, was able to discontinue four medications, rarely complains about severe pain in his joints, and is able to take regular walks without pain. Overall, his quality of life is significantly improved.

6.3.3 The unfulfilled wish to have children

Identifying and correcting micronutrient deficiencies can optimize the hormone metabolism's biochemical processes. In recent years, SALUTO tested 54 married couples, average age 36, who for years have desperately wanted to have children but so far decline hormone therapy from medical specialists.

The traditional blood tests done on the whole-blood level show no deficiencies. The results from cellular blood tests and those of the functional energy metabolism show a definite need for optimization in all areas, particularly severe deficiencies in the cellular micronutrient balance – specifically the individual amino acids (fig. 88). Moreover the 48-hour HRV analyses show a current dysregulation of the vegetative nervous system (fig. 90,91, 92).

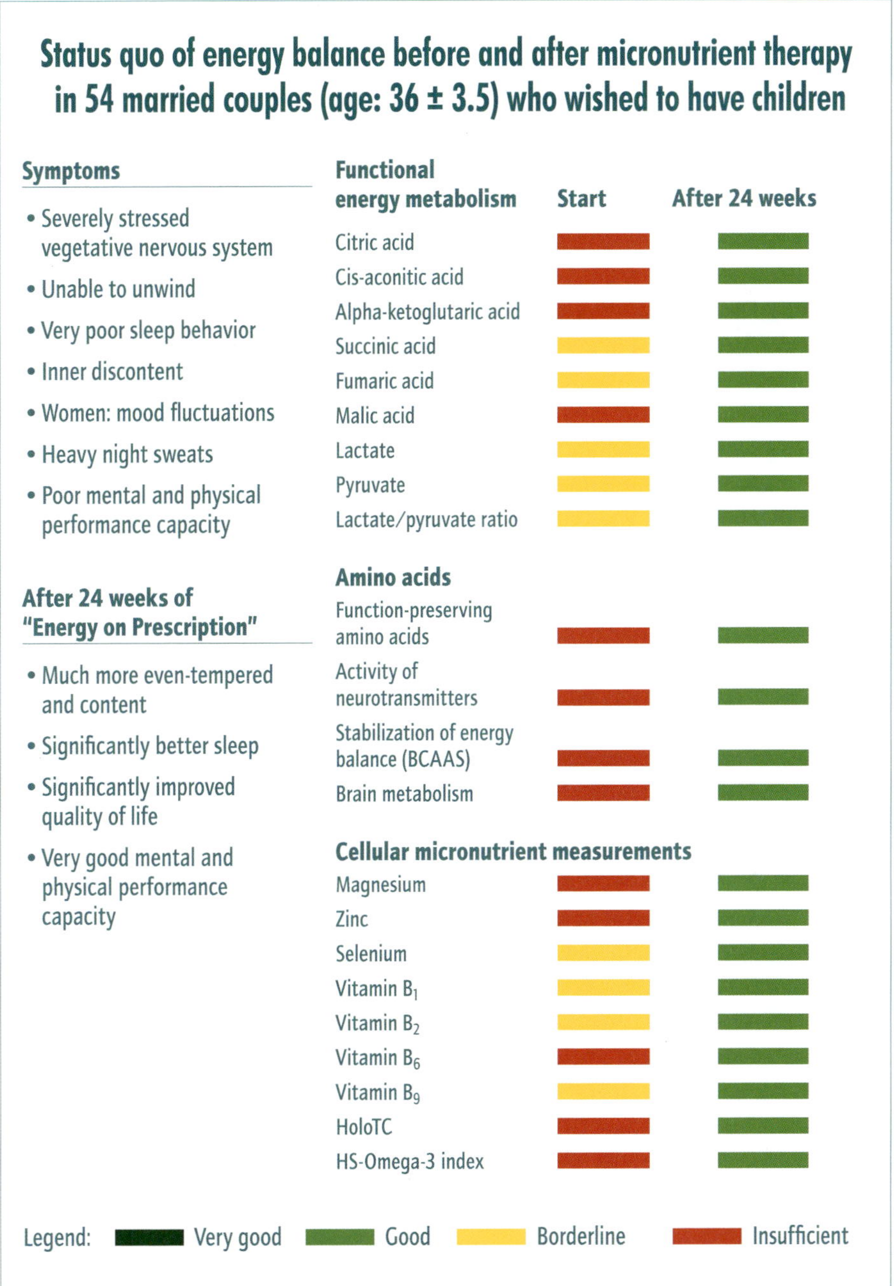

Fig. 88: Status quo of energy balance before and after micronutrient therapy in 54 married couples with the unfulfilled wish to have children

Furthermore, special blood tests show the following results:

- Significantly low HS-Omega-3 index of 3.9% FA;
- Intraerythrocytic deficiencies particularly of zinc, selenium, magnesium, Vitamins B_1, B_2, B_6 and B_9;
- In women frequently a functional iron deficiency – elevated sTfR value, ferritin 40 ng/ml;
- Striking deficiencies in special amino acids, arginine, tryptophan, phenylalanine, tyrosine, etc.;
- Thyroid hormone ranges TSH-basal < 1,3 µIU/ml and > 2,5 µIU/ml.

Based on the long-term experience of the author and his team, here the stress index and vegetative quotient in particular are high and the pNN50 is considerably low. These are unfavorable preconditions for fulfilling the mutual desire to have children. This is the case for men as well as women.

The mood of these couple is negative and understandably trends towards mental overload with the beginnings of exhaustion that initially could also be attributed to the previously measured insufficient thyroid hormone levels. In the women the thyroid antibodies TPO and TRAb show no signs of a current autoimmune disease such as Hashimoto thyroiditis, eliminating this as a possible cause.

The customized micronutrient intake results in a verifiable normalization or rather economization of the energy metabolism (fig. 89) and shows positive effects on the complex biochemical processes. If these can be successfully normalized, the preconditions for fulfilling the desire of having children are significantly improved. Particularly of note are the deficiencies in the area of the brain-activating amino acids tyrosine, tryptophan, and phenylalanine (fig. 88).

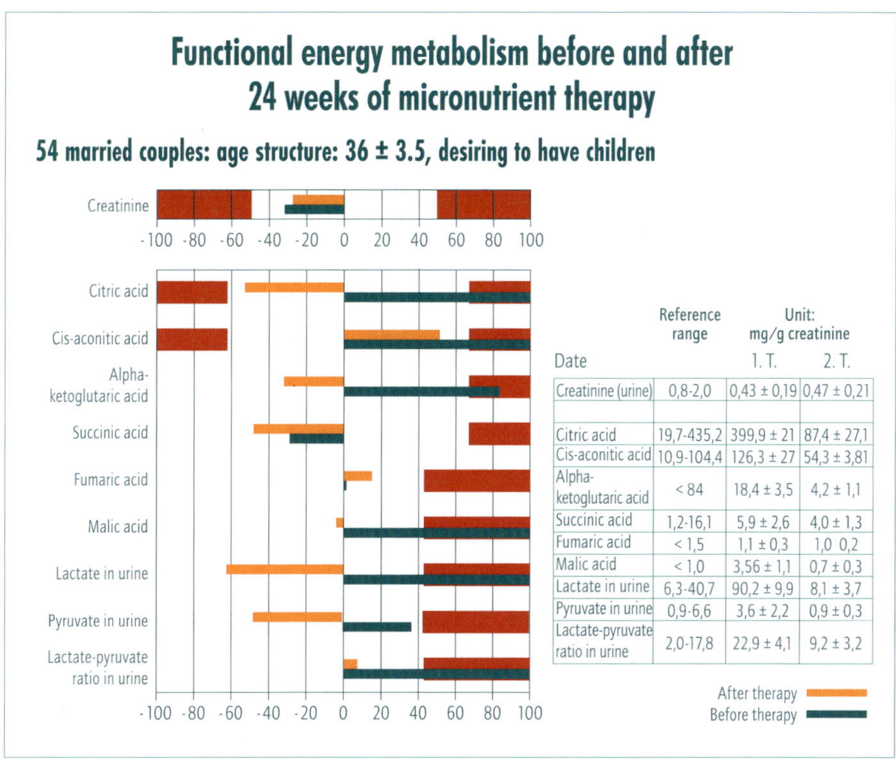

Date	Reference range	Unit: mg/g creatinine 1. T.	2. T.
Creatinine (urine)	0,8-2,0	0,43 ± 0,19	0,47 ± 0,21
Citric acid	19,7-435,2	399,9 ± 21	87,4 ± 27,1
Cis-aconitic acid	10,9-104,4	126,3 ± 27	54,3 ± 3,81
Alpha-ketoglutaric acid	< 84	18,4 ± 3,5	4,2 ± 1,1
Succinic acid	1,2-16,1	5,9 ± 2,6	4,0 ± 1,3
Fumaric acid	< 1,5	1,1 ± 0,3	1,0 0,2
Malic acid	< 1,0	3,56 ± 1,1	0,7 ± 0,3
Lactate in urine	6,3-40,7	90,2 ± 9,9	8,1 ± 3,7
Pyruvate in urine	0,9-6,6	3,6 ± 2,2	0,9 ± 0,3
Lactate-pyruvate ratio in urine	2,0-17,8	22,9 ± 4,1	9,2 ± 3,2

Fig. 89: Functional energy balance before and after micronutrient therapy in 54 married couples with the unfulfilled wish to have children

The results are matched to the database, making individually customized micronutrient formulations possible. The significantly low HS-Omega-3 index can verifiably decrease membrane elasticity in the individual cell structures and thereby impede a corresponding signaling cascade of neuronal structures. The importance of these individual amino acids to the entire brain and hormone metabolism has already been described in detail (➡ chapter 5.1).

When these deficiencies are corrected and the other lacking micronutrients are optimized, thyroid hormones verifiably normalize (tab. 21). The positive impact of the customized micronutrient intake on the normalization of thyroid hormones and the vegetative nervous system can be seen in couples whose desire to have children has not been fulfilled (fig. 90, 91, and 92). Here the correct dosage in particular is the measure of all things. The motto "more is more" is completely misplaced.

Tab. 21: Thyroid hormones and the balance of the vegetative nervous system before and after 24 weeks of customized micronutrient intake

	TSH- basal values bef. N = 29 0,79 ± 0,13	TSH- basal values bef. N = 25 3,61 ± 0,39	TSA- basal values after N = 29 1,91 ± 0,29	TSA- basal values after N = 25 2,1 ± 0,18
Stress index	410 ± 34,6	479 ± 48,1	177 ± 29,1	152 ± 20,1
pNN50	3,71 ± 2,4	4,99 ± 1,7	11,9 ± 3,21	13,1 ± 2,9
LF/HF ratio	6,92 ± 3,2	7,71 ± 3,6	1,98 ± 0,29	1,83 ± 0,17

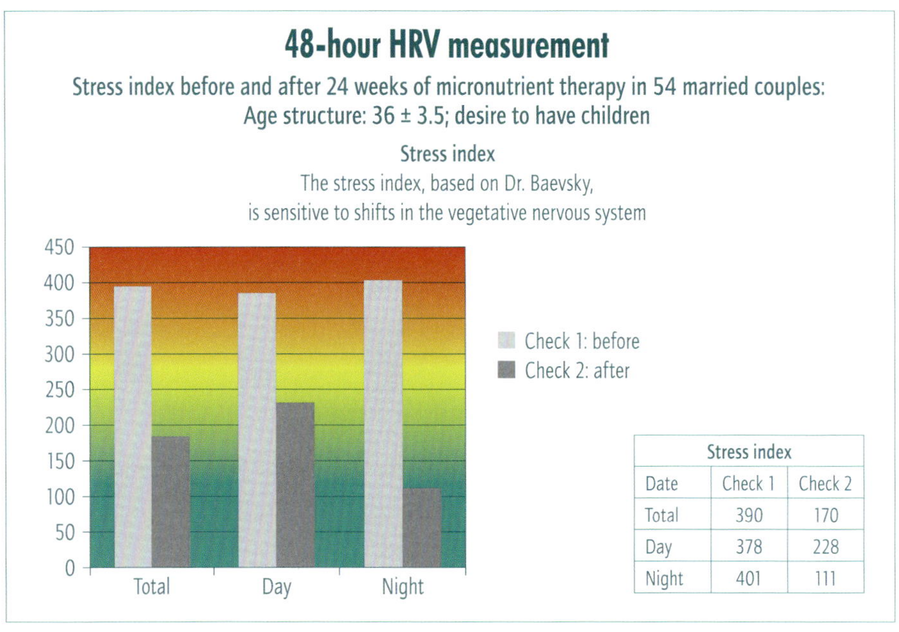

Stress index		
Date	Check 1	Check 2
Total	390	170
Day	378	228
Night	401	111

Fig. 90: Stress index before and after micronutrient therapy in 54 married couples with the unfulfilled wish to have children

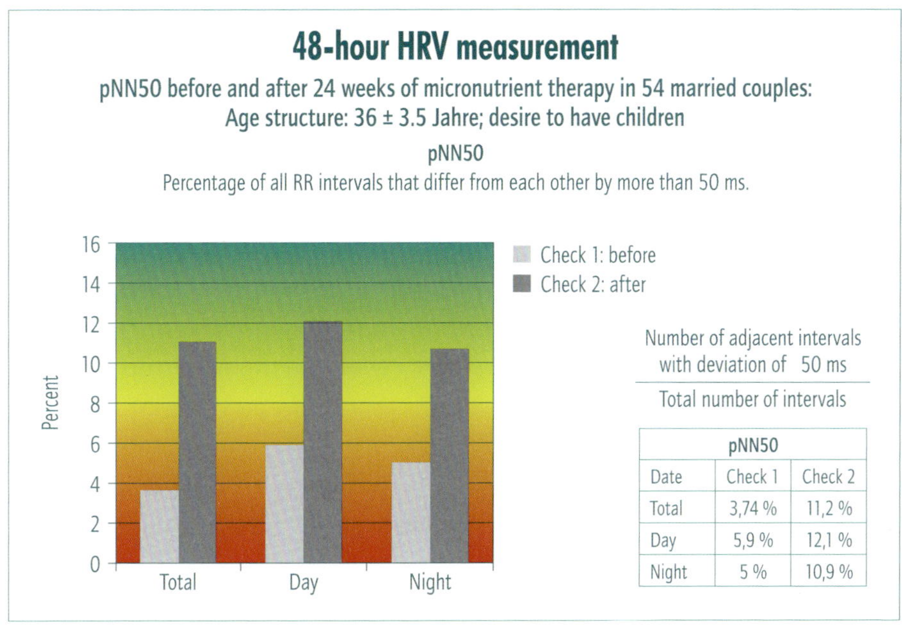

pNN50		
Date	Check 1	Check 2
Total	3,74 %	11,2 %
Day	5,9 %	12,1 %
Night	5 %	10,9 %

Fig. 91: pNN50 before and after micronutrient therapy in 54 married couples with the unfulfilled desire to have children

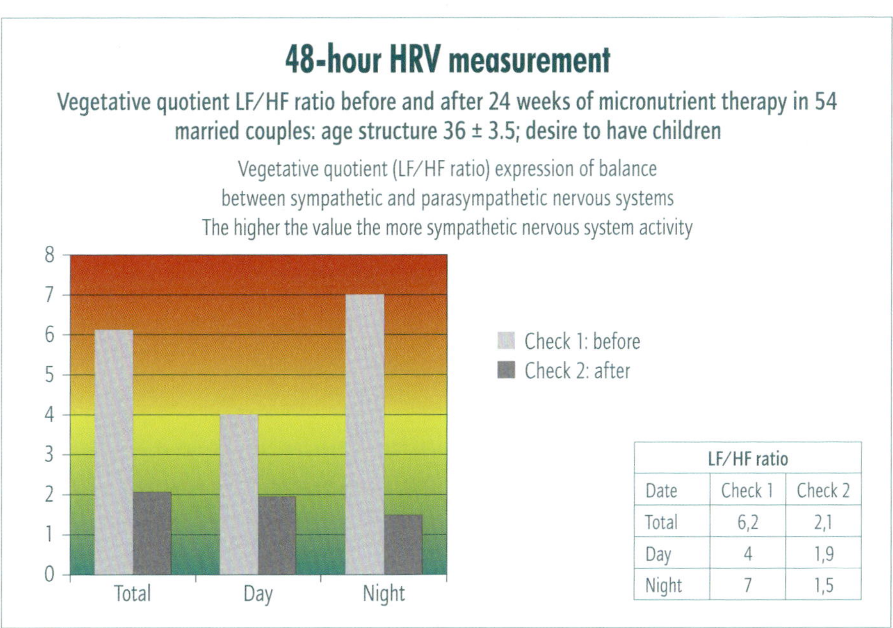

LF/HF ratio		
Date	Check 1	Check 2
Total	6,2	2,1
Day	4	1,9
Night	7	1,5

Fig. 92: Vegetative quotient before and after micronutrient therapy in 54 married couples with the unfulfilled desire to have children

CONCLUSION

For 48 of the 54 married couples their wish to have children came true within one year after customized micronutrient therapy and a targeted dietary change. Hence micronutrient supplementation is a very good formula for success.

The remaining six couples did not succeed in fulfilling their wish to have children due to medical reasons that made pregnancy impossible.

Hence our recommendation is: A targeted hormone treatment by medical specialists should be considered only after a customized micronutrient therapy and dietary change do not provide results.

6.3.4 Skin conditions (psoriasis, neurodermatitis, acne)

Results from recent years show direct links between complexion, microbiome, and mental stress, particularly with respect to psoriasis, neurodermatitis, and acne. A dietary change and the intake of lacking micronutrients can verifiably have a positive effect on the balance of the vegetative nervous system (fig. 93, 94, 95, 96). Everyday stressors can thereby be offset and the complexion improves.

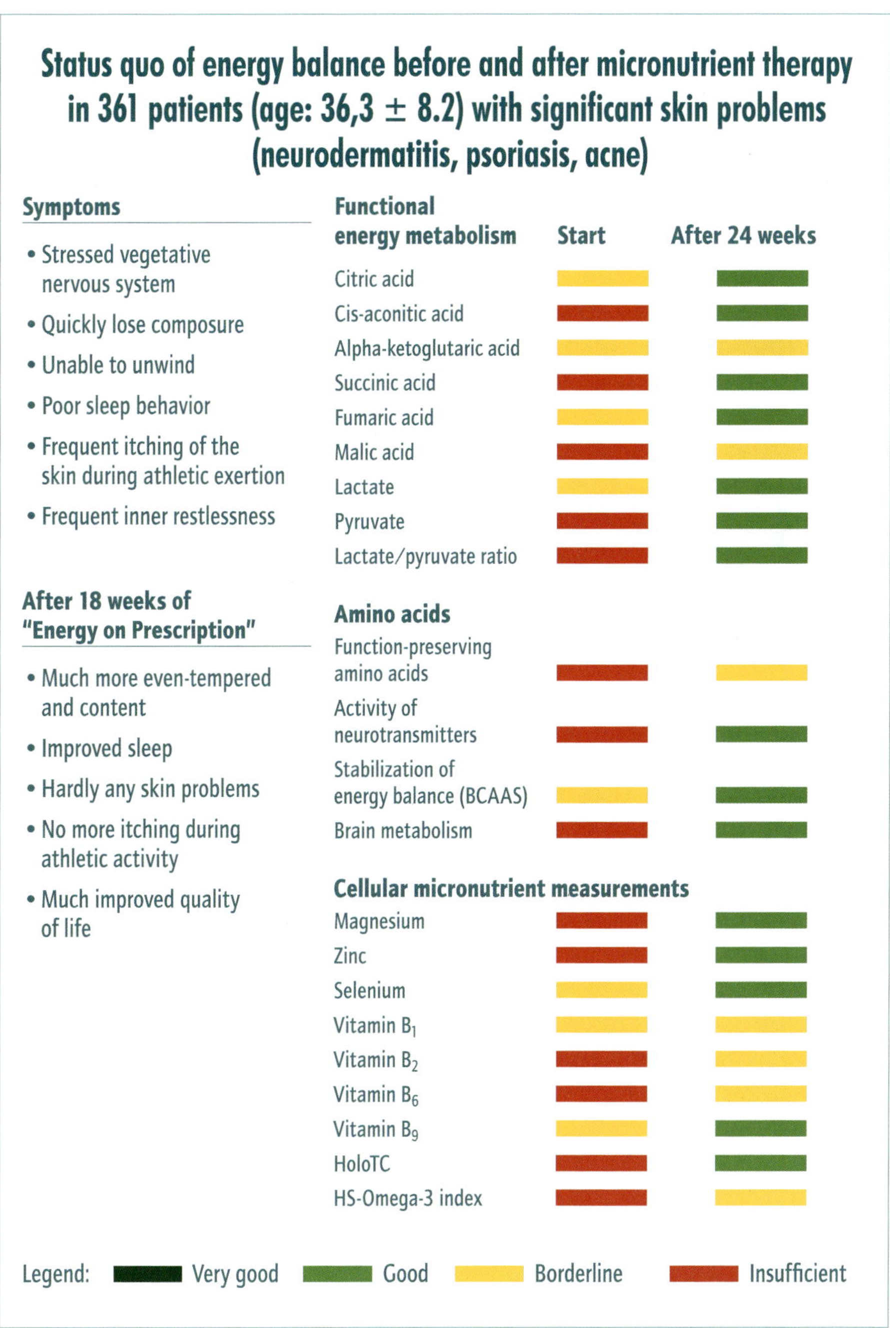

Fig. 93: Status quo of energy balance before and after micronutrient therapy in 361 dermatology patients

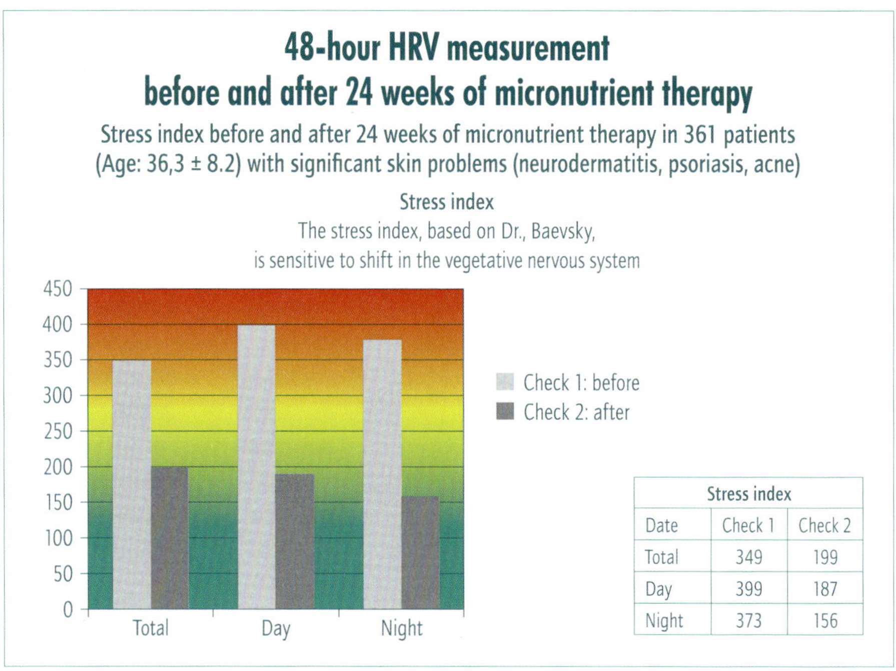

Stress index		
Date	Check 1	Check 2
Total	349	199
Day	399	187
Night	373	156

Fig. 94: Stress index before and after micronutrient therapy in 361 dermatology patients

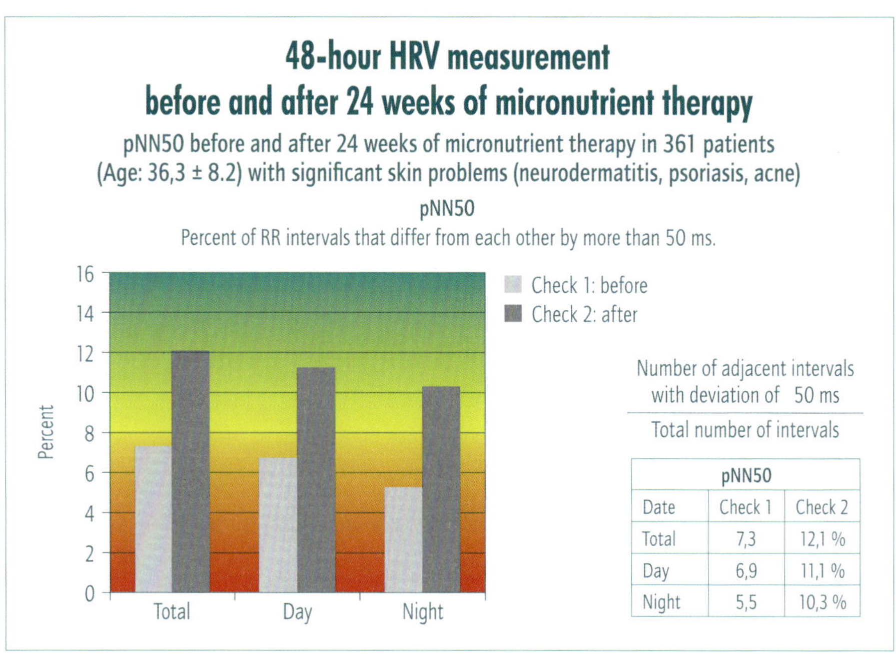

pNN50		
Date	Check 1	Check 2
Total	7,3	12,1 %
Day	6,9	11,1 %
Night	5,5	10,3 %

Fig. 95: pNN50 before and after micronutrient therapy in 361 dermatology patients

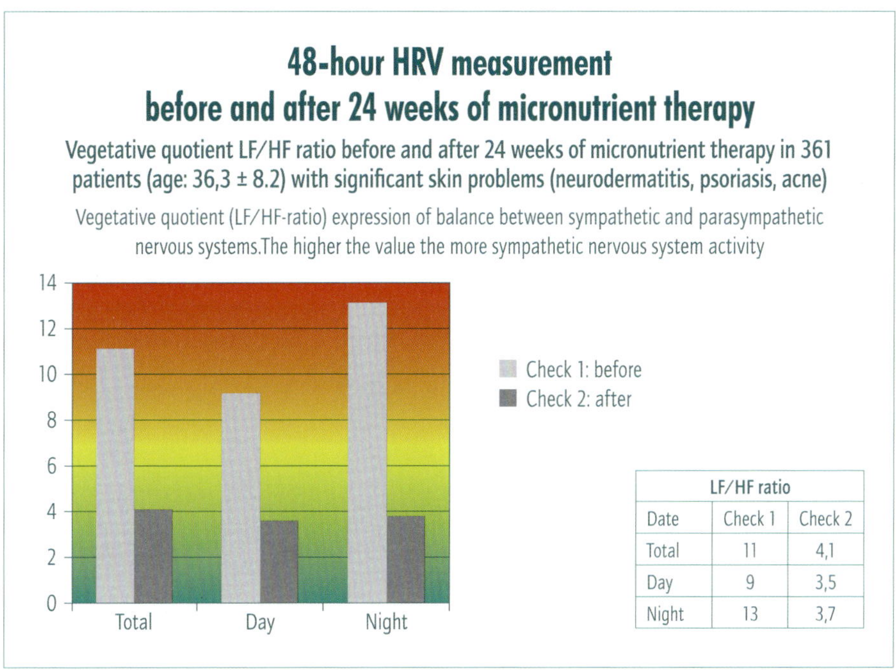

LF/HF ratio		
Date	Check 1	Check 2
Total	11	4,1
Day	9	3,5
Night	13	3,7

Fig. 96: Vegetative quotient before and after micronutrient therapy in 361 dermatology patients

The foundation for a good complexion is a well-functioning microbiome. Here the basic principles of Dr. Mosetter's Glycoplan have a very positive effect. An HS-Omega-3 index of > 12 % FA is important for a well-functioning microbiome, as it verifiably facilitates better and faster reduction of inflammatory reactions (silent inflammation).

Many evidence-based retrospective studies that were analyzed as part of SIP (study in practice) and master's theses, show an average HS-Omega-3 index of 3,15 ± 0,53. Cellular micronutrient concentrations also show severe deficiencies in magnesium, zinc, selenium, Vitamins B_1, B_2, B_6, and B_9 levels. This was also the case with individual amino acids such as phenylalanine, tryptophan, and tyrosine, which are particularly important for brain metabolism and a good mood.

6.3.4.1 Early detection and correction of micronutrient deficiencies improve the complexion

We did a retrospective evidence-based analysis (anthropometric data: tab. 22) of 100 patients with dermatological problems.

Tab. 22: Anthropometric data of 100 dermatological patients

	Without customized micronutrient intake with Glycoplan (n = 50)	**With** customized micronutrient intake with Glykoplan (n = 50)
Age (years)	21,23 ± 4,94	32,73 ± 6,94
Height (cm)	170,67 ± 6,25	179,97 ± 6,25
Weight (kg)	74,11 ± 4,72	78,83 ± 6,62

All of the 100 dermatology patients' diet was based on Dr. Mosetter's basic principles. The diagrams show the positive effect of micronutrients on the vegetative nervous system balance in 50 patients who were given an additional customized micronutrient formulation. Tab. 23 shows how the long-term glucose level, the so-called *HbA_{1C} value*, was reduced in both groups via dietary change, and also the effect of the targeted micronutrient intake.

Also, levels of the brain-activating amino acids in particular, phenylalanine and tryptophan, etc. (tab. 23 and fig. 97) rose with statistical significance after the micronutrient intake and resulted in improving the overall mood of the dermatology patients. Also of note is the initially considerably decreased Vitamin-D concentration (tab. 23 and fig. 97).

When dosing B vitamins it is often important to be mindful of the dosage since these patients can experience skin irritations with a dosage that is too high. We recommend a cautious approach.

Tab. 23: Blood test parameter of 100 dermatology patients with and without customized micronutrient intake

Parameter	Start (1. MTP)		After 12 weeks (2. MTP)		After 24 weeks (3. MTP)		Percent change (1. MTP → 3. MTP) p < 0,001 highly significant		Comparison test 1. To 3. MTP between the two groups
	without	with	without	with	without	with	without	with	
HS-Omega-3 Index (% FS)	3,31 ± 0,51	2,91 ± 0,33	3,39 ± 0,41	9,79 ± 0,59	3,21 ± 0,25	9,38 ± 0,35	2,56	242,96	p < 0,001
HbA_{1c} (%, mM/M)	5,51 ± 0,19	5,72 ± 0,27	4,89 ± 0,38	4,97 ± 0,33	4,77 ± 0,29	4,89 ± 0,25	- 13,43	-14,51	p < 0,001
Vitamin D (nmol/l)	23,93 ± 7,6	26,51 ± 6,9	21,6 ± 6,4	124,45 ± 5,0	23,9 ± 6,3	131,37 ± 7,9	-0,13	366,56	p < 0,001
Tryptophan (mg/dl)	1,19 ± 0,15	1,05 ± 0,2	1,25 ± 0,11	1,71 ± 0,29	1,22 ± 0,13	1,92 ± 0,23	2,52	39,56	p < 0,001
Phenylalanine (mg/dl)	1,17 ± 0,13	1,28 ± 0,1	1,86 ± 0,14	2,63 ± 0,14	1,36 ± 0,18	2,71 ± 0,13	15,24	111,72	p < 0,001
Vitamin B_9 (µg/Ery.)	534,1 ± 77,3	558,1 ±	551,3 ± 59,7	998,3 ± 98,1	549,2 ± 67,8	1.21 ± 101,3	2,83	83,60	p < 0,001

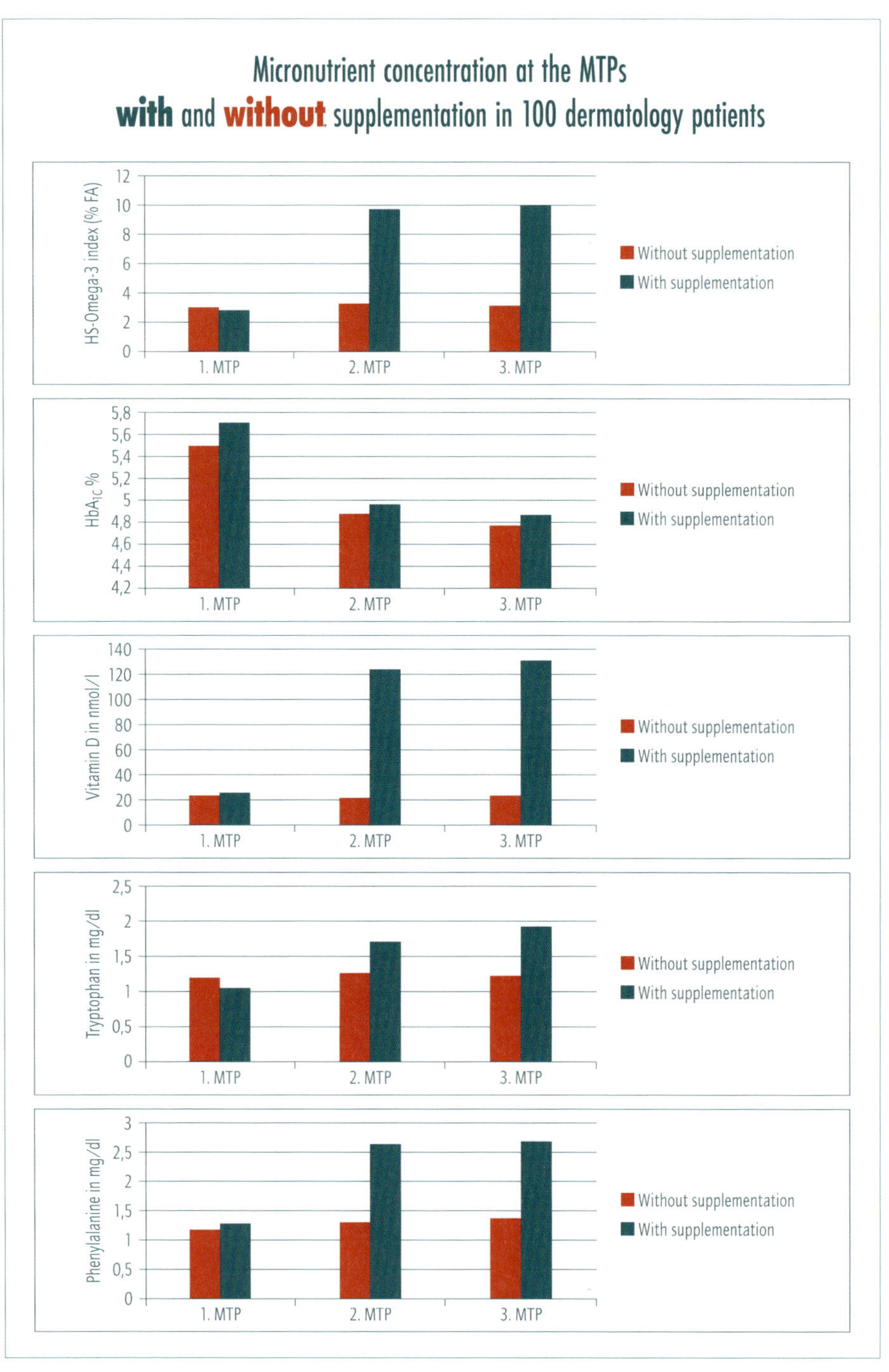

Fig. 97: Micronutrient concentrations at three measurement time points with and without micronutrient supplementation in 100 dermatology patients

Fig. 98 and fig. 99 show the positive effects of micronutrient supplementation on the vegetative nervous system and thyroid hormones.

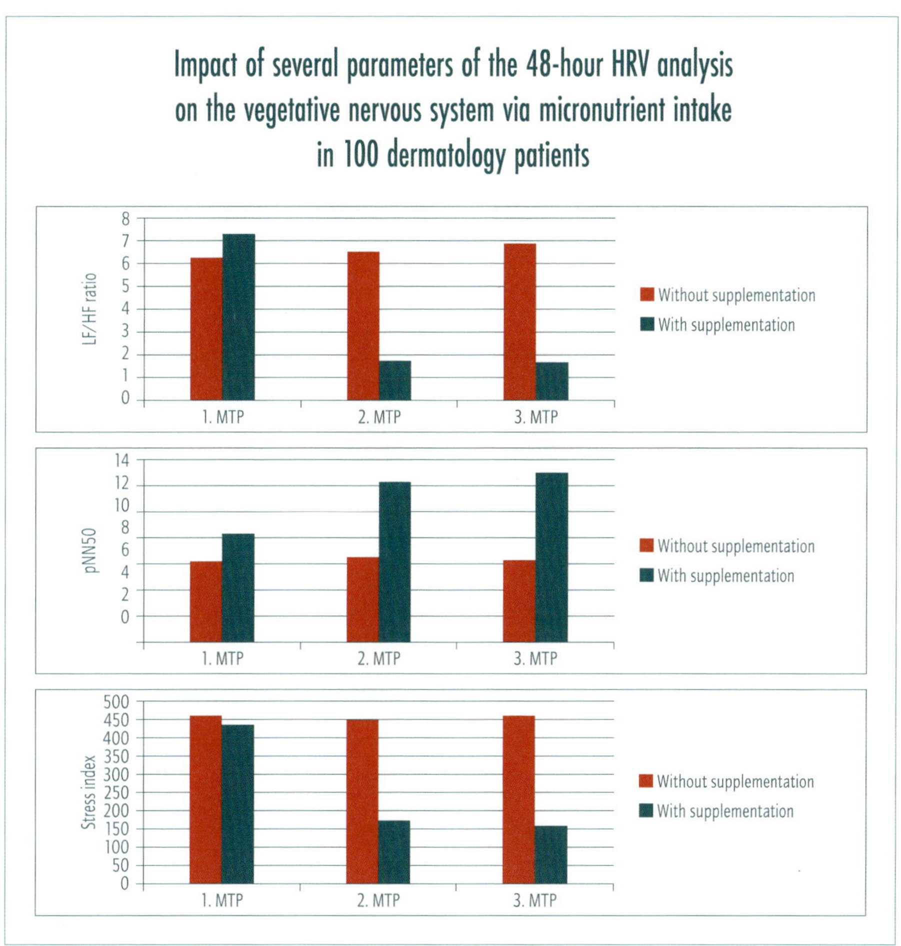

Fig. 98: Parameters of the 48-hour HRV analysis at three measurement time points with and without micronutrient supplementation in 100 dermatology patients

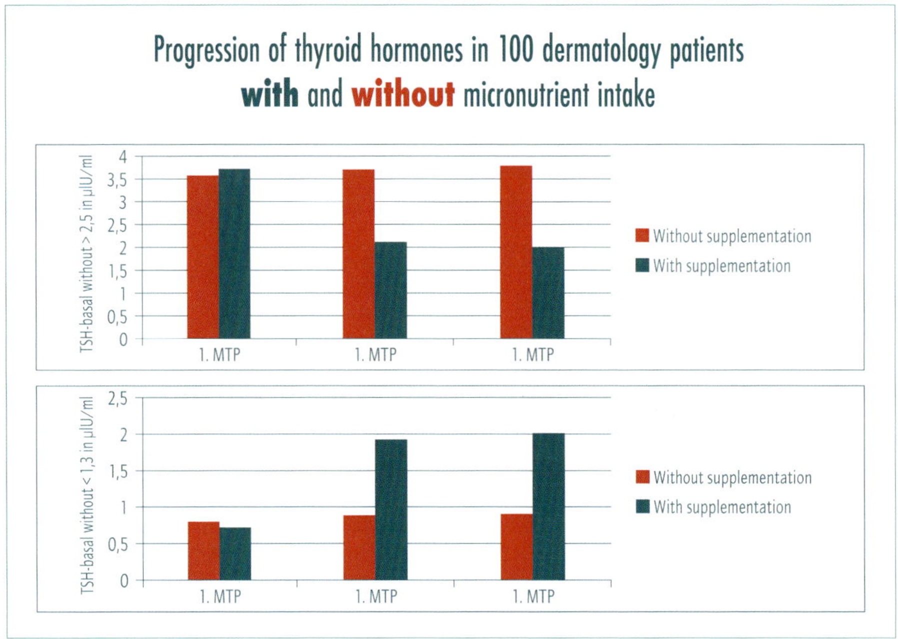

Fig. 99: Thyroid hormone concentration at three measurement time points with and without micronutrient supplementation in 100 dermatology patients

Overall, the analysis showed

- ... that a thyroid imbalance results in decreased HRV and thus stresses the body.
- ... that the TSH basal value and the HRV parameters stress index, vegetative quotient (LF/HF ratio), and pNN50 are linked.
- ... that the targeted and customized micronutrient formulation can facilitate regulation and normalization of thyroid hormones and the vegetative nervous system (sympathetic and parasympathetic nervous systems) and thereby leads to an even complexion.
- ... that symptoms such as poor sleep behavior, increased fatigue, diminished mental performance, and inner restlessness increasingly arise from a suboptimal, but according to conventional medicine, not yet pathological TSH basal value.
- ... that a targeted micronutrient intake verifiably reduces stressors (rise in pNN50, stress index, LF/HF ratio as vegetative quotient; fig. 98) and the complexion improves remarkably.
- ... that a dietary change alone will not normalize thyroid hormones (fig. 99).
- ... that previous reference values and an evaluation of thyroid hormones should be called into question.
- ... that there is a statistically significant positive or negative link to point T_2 and T_3, whereby the correlation coefficient lies between 0.89 and 0.91.

6.3.5 Chronic fatigue syndrome, prevailing depressive mood

The results from our database show that chronic fatigue syndrome as well as a prevailing depressive mood are often accompanied by changes in micronutrients (fig. 100). This is plausible since brain metabolism biochemistry is subject to micronutrients, which can manifest itself in changes to the stress index, pNN50, and the vegetative quotient (fig. 101, 102, 103).

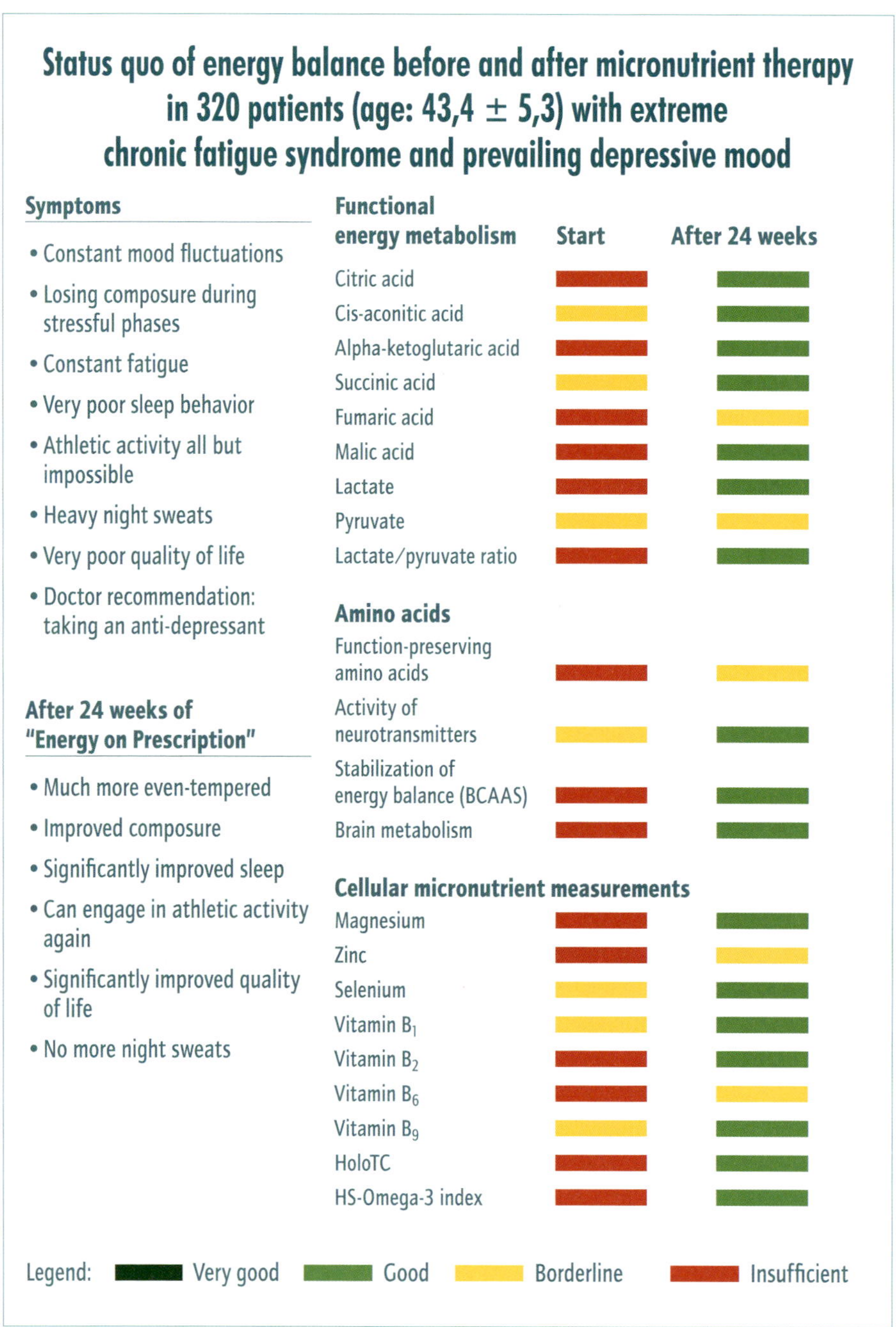

Fig. 100: Status quo of energy balance before and after micronutrient therapy in 320 patients with extreme chronic fatigue syndrome and prevailing depressive mood

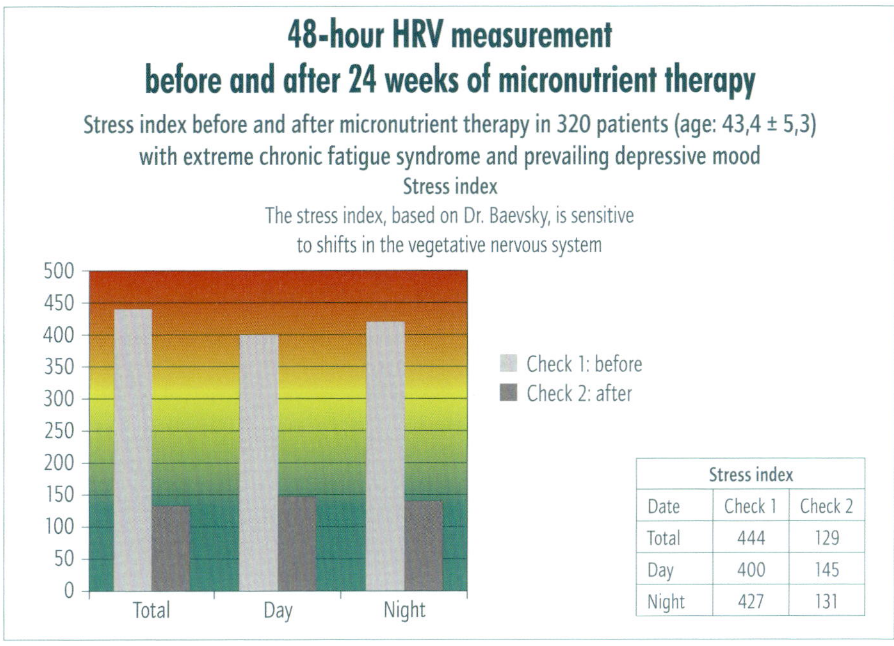

Stress index		
Date	Check 1	Check 2
Total	444	129
Day	400	145
Night	427	131

Fig. 101: Stress index before and after micronutrient therapy in 320 patients with extreme chronic fatigue syndrome and prevailing depressive mood

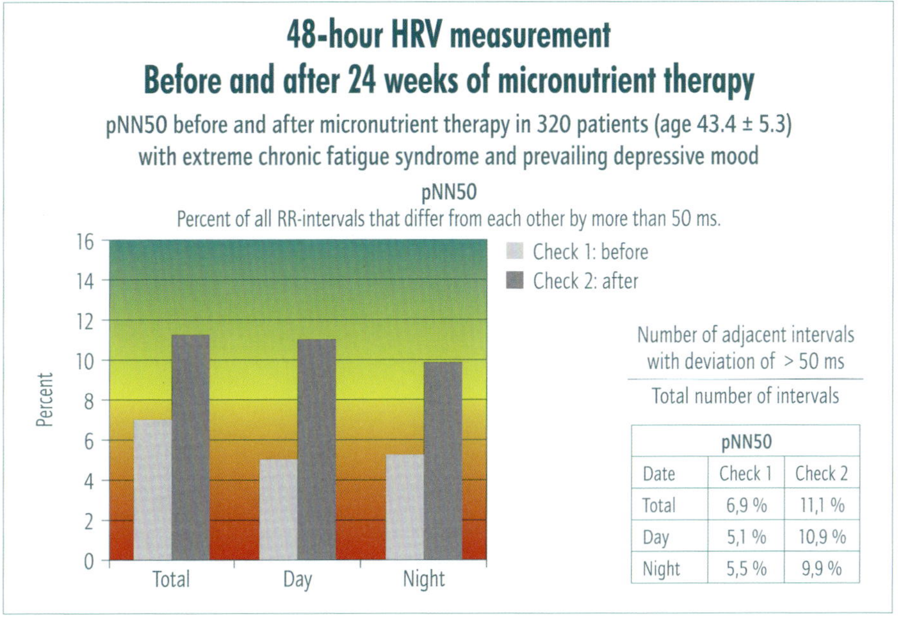

pNN50		
Date	Check 1	Check 2
Total	6,9 %	11,1 %
Day	5,1 %	10,9 %
Night	5,5 %	9,9 %

Fig. 102: pNN50 before and after micronutrient therapy in 320 patients with extreme chronic fatigue syndrome and prevailing depressive mood

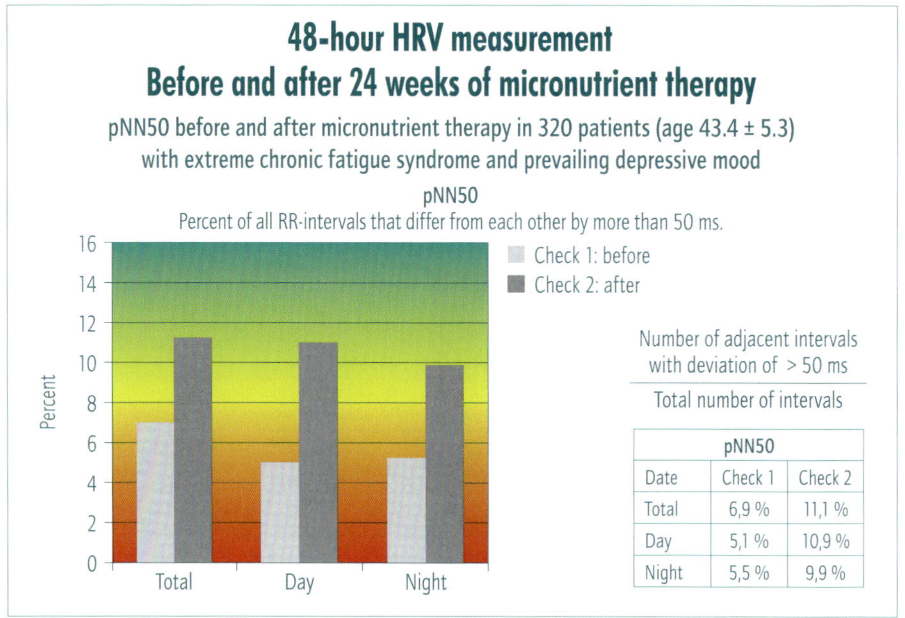

pNN50		
Date	Check 1	Check 2
Total	6,9 %	11,1 %
Day	5,1 %	10,9 %
Night	5,5 %	9,9 %

Fig. 102: pNN50 before and after micronutrient therapy in 320 patients with extreme chronic fatigue syndrome and prevailing depressive mood

A good mood, peak performance, creativity all depend on a sufficient supply of amino acids that influence the functional energy metabolism. The particular impact of diet on our sense of wellbeing is explicitly delineated in chapter 4.1. Mental and physical stress can significantly increase the demand.

Amino acids as nutrients have a direct impact on your personal wellbeing. By choosing the right foods you can personally affect your status and achieve a "good mood". Euphoria and optimism are reflections of your amino acids. An optimal supply of the accompanying elementary amino acids will quickly let sadness and a bad mood fade away. A signaling effect on neuronal structures cannot take place without an adequate supply of B vitamins and Omega-3 fatty acids.

The basic principles of an optimized diet are the foundation for "feeling good".

6.3.5.1 Case study of a 42-year-old female

(This patient is not the same person as the "sample patient" in chapter 6.2.1.2.)

Over the course of a year, the patient experienced increasingly severe malaise, extreme mood swings, and even depression-like symptoms. She felt like she no longer knew who she was. She is a 42-year-old woman, height 167 cm, weight 56 kg, a successful executive, who does endurance training 3 x week for 30-40 minutes.

But now she suffers from chronic fatigue syndrome (exhaustion), frequent inner restlessness, muscle tension, trouble falling and staying asleep. She has been receiving psychosomatic-disorder treatment for the past three months and now has been told to take an antidepressant.

42-year-old woman with extreme chronic fatigue syndrome and prevailing depressive mood

TSH-basal value	0,84 µIU/ml	Sodium	141 nmol/l
Ferritin	15,1 ng/ml	Calcium	2,54 nmol/l
Uric acid	5,6 mg/d	Hematocrit	56,4 %
Chol	190 mg/dl	Leucocytes	7,80 Tsd/µl
HDL	62 mg/dl	Erythrocytes	4,81 Mio/µl
LDL	124 mg/dl	Hemoglobin	12,1 g/dl
HbA_{1c}	5,5 % (36,7 mM/M)	MCV	96,2 fl
CRP	< 0,30 mg/dl	MCH	32,5 pg
GOT	48 U/l	MCHC	31,1 g/dl
GPT	18 U/l	Thrombocytes	224 Tsd/µl
GGT	11 U/l		
CK	142 U/l		
Alkaline phosphatase	60 U/l	Testosterone: 0,06 ng/ml	
Total bilirubin	0,32 mg/dl	Hormone status normal	
Iron	149 µg/dl	I-FABP: 2.361 pg/ml (389-2.129)	
Transferrin receptor soluble	1,73 mg/l	Thyroid antibodies TPO are suspicious	
Urea	33 mg/dl	Takes L-thyroxine 50	

Fig. 104: Lab results from a 42-year-old patient

Cardiological and general internal-medicine testing did not reveal any abnormalities. Menstrual cycle and gynecological hormone status are normal and do not suggest the beginning of menopause. She takes 50 mg of L-thyroxine due to an autoimmune disorder of the thyroid (Hashimoto thyroiditis). Fig. 104 shows the lab results.

The intracellular analysis of magnesium, zinc, selenium, Vitamin-B_9 and HoloTC (active Vitamin-B_{12}) show decreased concentrations. The special amino acids (arginine, phenylalanine, tryptophan, tyrosine, etc.) show significant need of optimization.

The intestinal fatty-acid binding protein (I-FABP) currently seems the best-suited marker for a disrupted intestinal barrier and in this case is significantly elevated. I-FABP exists exclusively in the cytoplasm of enterocytes (100% specific) and here plays a role in the fatty-acid metabolism. If the enterocyte is damaged the I-FABP is released into the circulation and can be measured in serum. Compared to Zonulin, the I-FABP level is more significant.

For this reason, the 42-year-old woman will first change her diet for 3 weeks before starting the micronutrient formulation. After 24 weeks, the initially 5.5% elevated HbA_{1C} (36.7 mM/M) dropped to 5.0 (33.4 mM/M) and thus created significantly better conditions for the functional energy metabolism.

The significantly low ferritin concentration (15.1 ng/ml) and the elevated sTfR (transferrin receptor; 1.73 mg/l) showed a definite deficiency in functional iron. The woman received a one-time iron infusion with Ferinject 50 mg vials (equals 200 mg of iron) as a 30-minute infusion since an oral intake does not make sense during this phase due to the currently disrupted intestinal barrier, and would take too long.

After the completed follow-up exam three weeks later, the 42-year-old woman regularly took ferro sanol duodenal three times a week on Monday, Wednesday, and Friday, with the intention of maintaining the ferritin concentration. After 24 weeks, the regular blood tests showed a ferritin level of 71 ng/ml (optimal: > 60 and < 150 mg/ml).

42-year-old woman with extreme chronic fatigue syndrome and prevailing depressive mood

Agent Vitamins	Daily dose	Agent Vitamins	Daily dose
Vitamins		**Trace elements**	
Vitamin A (retinol)	1 mg	Chrome	250 µg
Vitamin B_1 (thiamine)	50 mg	Selenium	185 µg
Vitamin B_2 (riboflavin)	50 mg	Zinc	48 mg
Niacin (Vitamin B_3)	70 mg		
Vitamin B_6 (pyridoxin)	90 mg	**Minerals**	
Vitamin B_{12} (cyanocobalamin)	1.250 µg	Calcium	400 mg
Vitamin C (ascorbic acid)	2.000 mg	Silicon	420 mg
Vitamin D	160 µg		
Natural Vitamin E	150 mg	**Quasi-vitamins**	
Of that alpha tocopherol	107,5 mg	Choline	160 mg
Gama tocopherol	42,5 mg	Coenyme Q_{10}	120 mg
Natural carotenoids	8 mg	Inositol	120 mg
Of that alpha carotin	70 µg	PABA	60 mg
Beta carotin	1,9 mg		
Cryptoxanthin	15 µg	**Plant extracts**	
Lutein	6 mg	Citrus flavonoids	275,9 mg
Zeaxanthin	15 mg		
Biotin (Vitamin H)	160 µg		
Folic acid (Vitamin B_9)	1.200 µg		
Pantothenic acid	120 mg		
Additional amino acids in special cases			
Phenylalanine	1.000 mg		

One-time iron infusion ferinject
Oral intake after three weeks 3 x week,
Mo – Wed – Fri ferro sanol duodenal 100 mg

Daily dosage
14 ml morning and midday
After 12 weeks only 14 ml in the morning

Additionally

- Arginine in powder form 3 g/3 g/0
- Magnesium bisglycinate 100 mg capsule 2/2/2
- 2 Tbs Norsan Omega-3 oil (1 Tbs = 1.700 mg EPA/DHA), 2/0/0
- 5-HTP (hydroxytryptophan)Griffonia 100 mg 0/0/2 1 hour before bed

Fig. 105: Micronutrient formulation of a 42-year-old female patient

The initially measured TSH basal value of 0.84 µIU/l while taking L-thyroxine with an existing autoimmune disorder of Hashimoto thyroiditis results in an elevated stress index, decreased pNN50, and an elevated vegetative quotient LF/HF ratio. After 24 weeks, decreasing the medication to L-thyroxine 25 and the targeted intake of lacking micronutrients resulted in a rise of the TSH-values to 1.9 µIU/l. This is where we find the sense-of-wellbeing effect.

The physician recommended use of an antidepressant is entirely the wrong approach for this patient. It would only treat the symptom and not the actual cause. The simultaneous intake of the lacking amino acids phenylalanine, tryptophan, and tyrosine that are critical to the brain metabolism, has already achieved a very good balance of the vegetative nervous system after just a few weeks. The progression of vegetative nervous system parameters shows the extraordinary effect of micronutrient therapy. This also proves to be true for the functional energy metabolism (fig. 106).

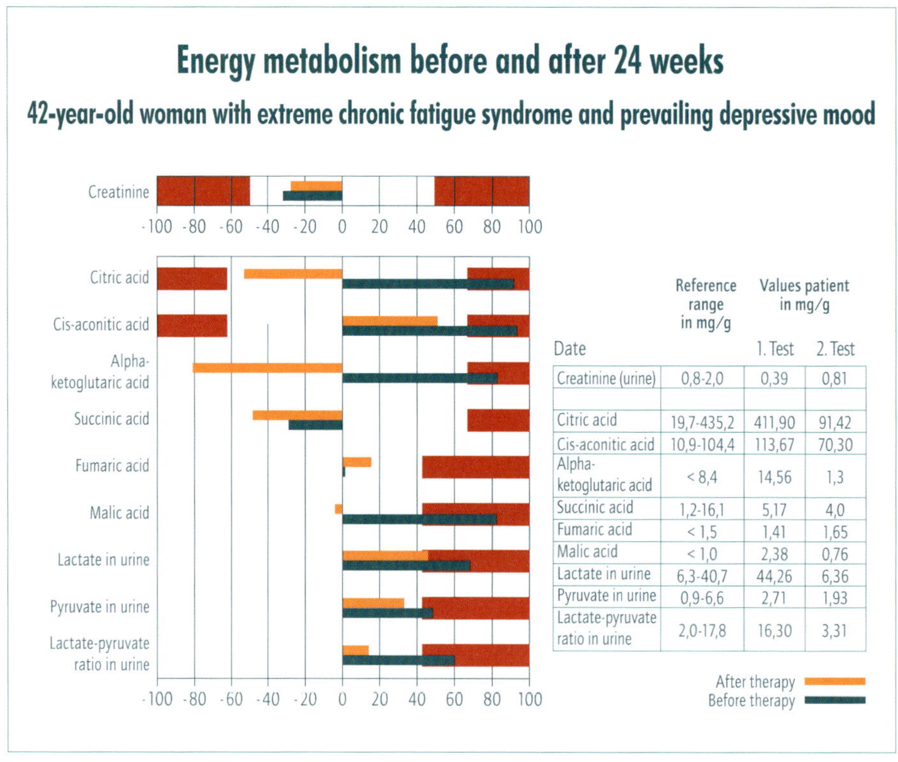

	Reference range in mg/g	Values patient in mg/g	
Date		1. Test	2. Test
Creatinine (urine)	0,8-2,0	0,39	0,81
Citric acid	19,7-435,2	411,90	91,42
Cis-aconitic acid	10,9-104,4	113,67	70,30
Alpha-ketoglutaric acid	< 8,4	14,56	1,3
Succinic acid	1,2-16,1	5,17	4,0
Fumaric acid	< 1,5	1,41	1,65
Malic acid	< 1,0	2,38	0,76
Lactate in urine	6,3-40,7	44,26	6,36
Pyruvate in urine	0,9-6,6	2,71	1,93
Lactate-pyruvate ratio in urine	2,0-17,8	16,30	3,31

Fig. 106: Functional energy metabolism before and after micronutrient therapy

Here is an **important tip**: L-tryptophan intake in the form of a protein powder is contraindicated for this patient as this results in neurotoxic reactions (kynurenine).

This can further compound the existing problem. Due to the active autoimmune disease of the thyroid the patient should only receive the amino acid 5-HTP (fig. 105).

A replenishing phase of 12 weeks is indicated for this patient, during which the cellular structures are filled. After that the micronutrient formulation is reduced so the functional energy metabolism can further normalize or economize itself.

42-year-old woman with extreme chronic fatigue syndrome and prevailing depressive mood

Before

- Chronic fatigue syndrome (state of exhaustion)
- Increasing depressive mood
- Frequent inner restlessness and very stressed vegetative nervous system
- Significant muscle tension
- Inability to fall and stay asleep
- Doctor's recommendation: psychosomatic treatment needed
- And current recommendation: use of an antidepressant
- Increasing hair loss

After 24 weeks of micronutrient therapy and dietary change

- Very good mental and physical performance ability
- Meanwhile good composure
- No muscle tension
- 48-hour HRV measurement shows a balanced vegetative nervous system
- Good ability to fall and stay asleep
- Feels great after athletic exertion
- Stabilization of hair structure

Fig. 107: Subjective state of health of a 42-year-old female patient

Fig. 107 shows the effect of the functional energy metabolism with the customized micronutrient formulation and the positive biochemical changes that have led to a completely improved mental and physical performance in the 42-year-old woman. The 48-hour HRV measurements before and after the 24-week micronutrient therapy show a definite lowering of the stress index (fig. 108), the vegetative quotient LF/HF ratio (fig. 110), and a rise in pNN50 (fig. 109).

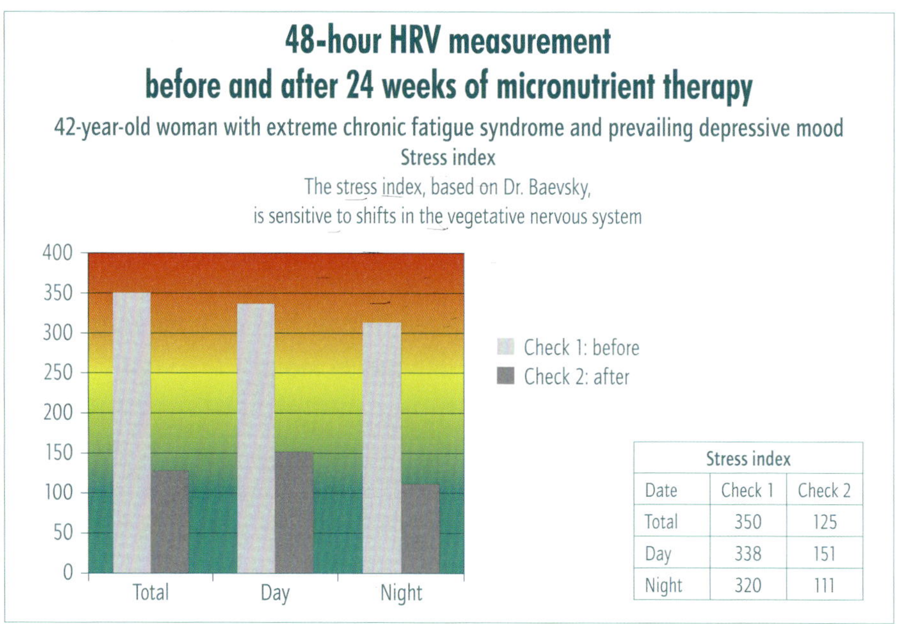

Stress index		
Date	Check 1	Check 2
Total	350	125
Day	338	151
Night	320	111

Fig. 108: Stress index of the 42-year-old female patient

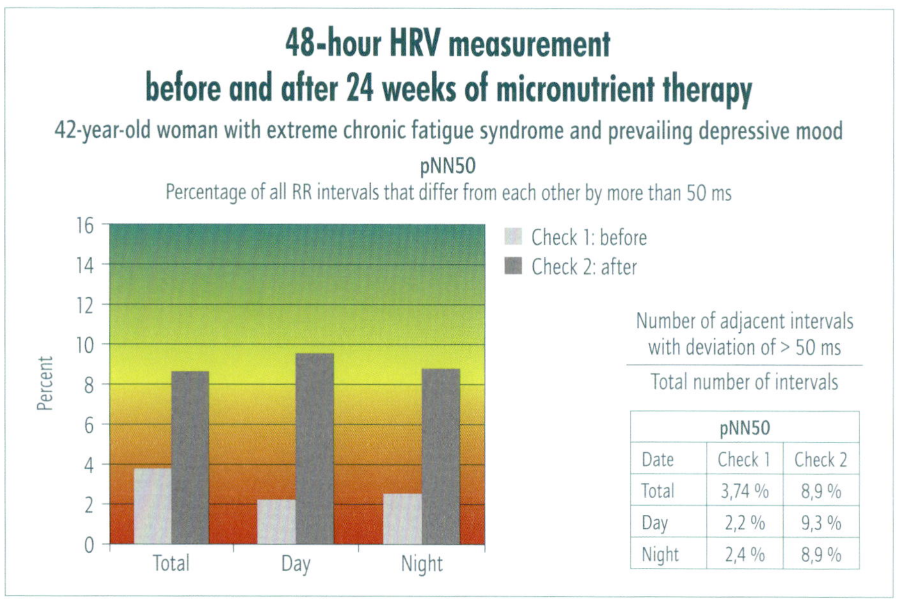

pNN50		
Date	Check 1	Check 2
Total	3,74 %	8,9 %
Day	2,2 %	9,3 %
Night	2,4 %	8,9 %

Fig. 109: pNN50 of the 42-year-old female patient

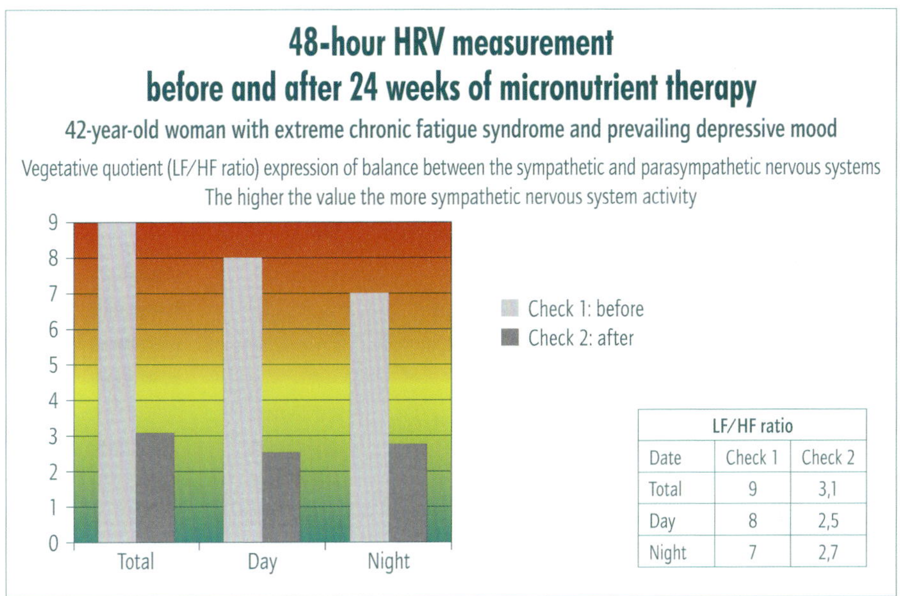

LF/HF ratio		
Date	Check 1	Check 2
Total	9	3,1
Day	8	2,5
Night	7	2,7

Fig. 110: Vegetative quotient in the 42-year-old woman

6.3.6 Menopausal symptoms in women

In recent years, the author and his team have been able to ascertain a depleted "energy balance" in more than 850 women within the ages 46-52 with severe menopausal symptoms, in spite of traditional medical tests on the serum level not showing any deficiencies that could verifiably cause many disorders. 420 women aged 48,4 ± 5,3 underwent investigative testing.

The functional energy metabolism (fig. 111) shows severe deficiencies and significantly impaired activity of certain enzymes. The TSH-basal values are at levels of < 1,3 µIU/ml and > 2,5 µIU/ml. A "sense of wellbeing" is no longer possible in those ranges.

Of the tested women 10% show the beginnings of an autoimmune disease of the thyroid as the TPO or TRAb levels are elevated. The tested women report various symptoms, from the onset of insomnia to sweating, inner restlessness, to weight gain. They quickly lose their composure during stressful phases and describe a poor quality of life.

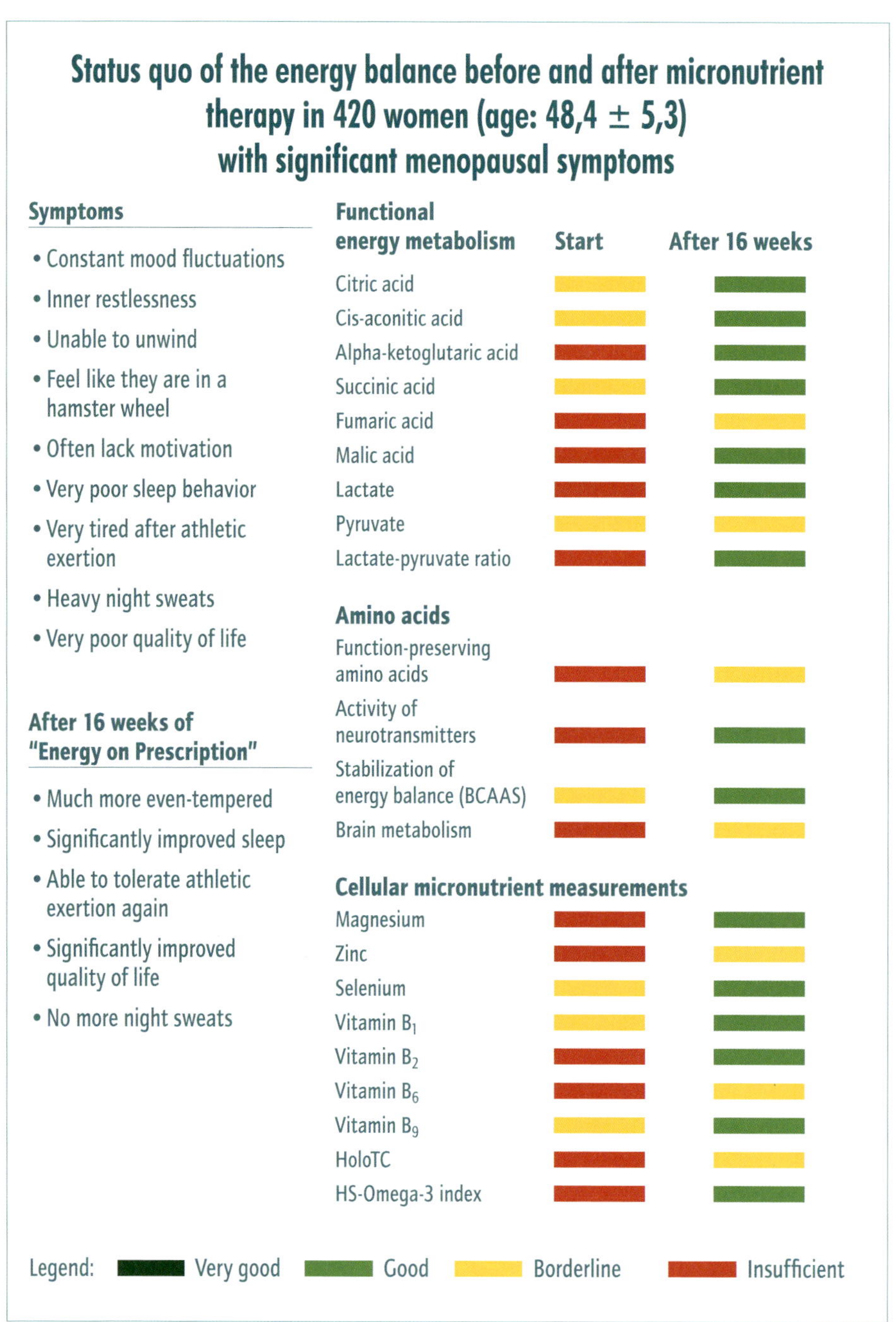

Fig. 111: Status quo of the energy balance before and after micronutrient therapy in 420 women with severe menopausal symptoms

Menopausal symptoms in women can also be significantly reduced with an intake of lacking micronutrients (fig. 112, 113, 114). This aspect should be taken into consideration during gynecological exams, particularly because hormone therapy is accompanied by an increased risk of cancer.

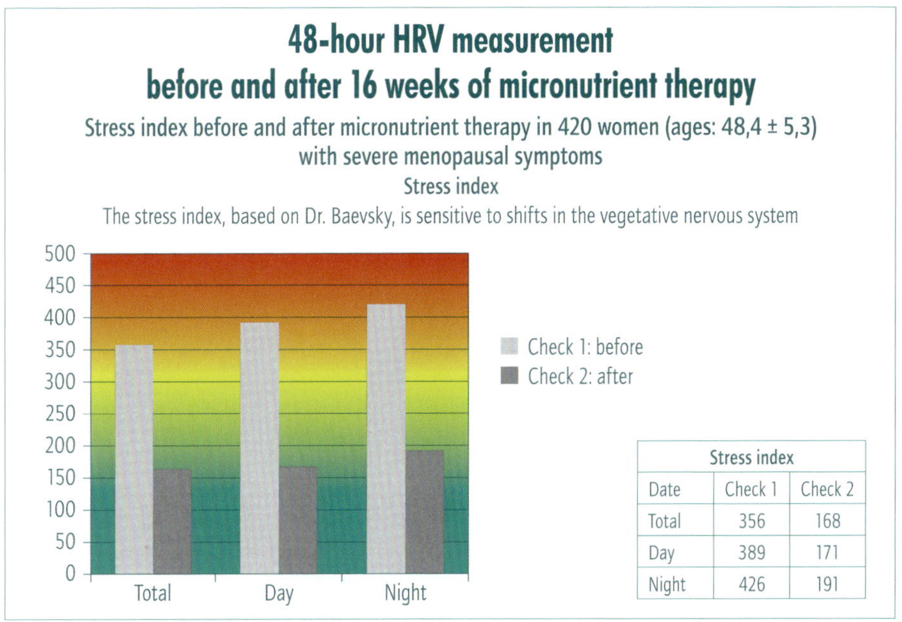

Stress index		
Date	Check 1	Check 2
Total	356	168
Day	389	171
Night	426	191

Fig. 112: Stress index before and after micronutrient therapy in 420 women with severe menopausal symptoms

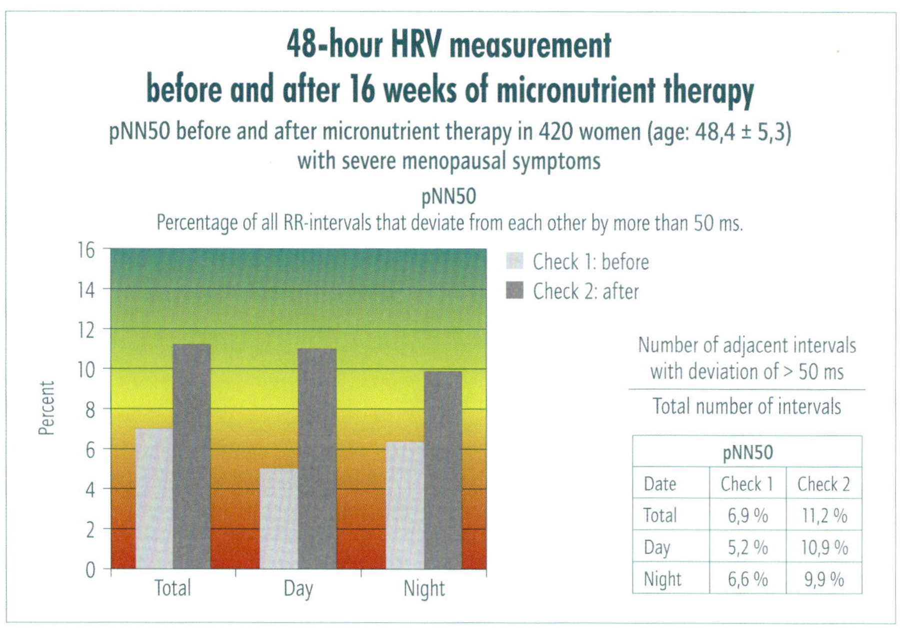

pNN50		
Date	Check 1	Check 2
Total	6,9 %	11,2 %
Day	5,2 %	10,9 %
Night	6,6 %	9,9 %

Fig. 113: pNN50 before and after micronutrient therapy in 420 women with severe menopausal symptoms

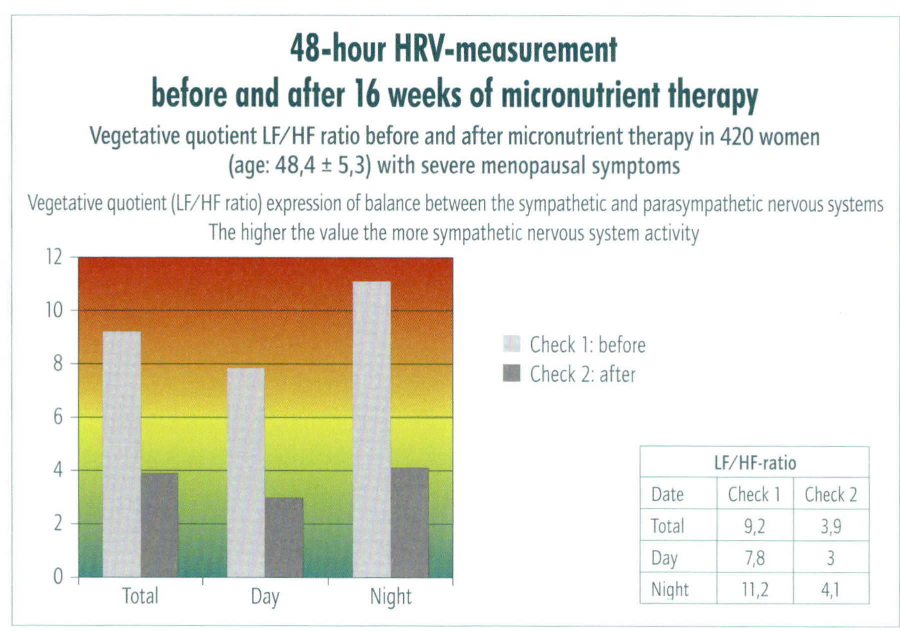

LF/HF-ratio		
Date	Check 1	Check 2
Total	9,2	3,9
Day	7,8	3
Night	11,2	4,1

Fig. 114: Vegetative quotient before and after micronutrient therapy in 420 women with severe menopausal symptoms

The following example of a 46-year-old female teacher shows how successful the customized micronutrient therapy is in helping women regain quality of life by ascertaining existing biochemical disruptions early and correcting them.

6.3.6.1 Case study of a 46-year-old woman

The woman – height 173 cam, weight 60 kg – complained of inner restlessness, sleep disturbances, also frequent lack of motivation, exhaustion, and fatigue. She had a weak immune system. She also reported having a pollen/dust allergy. Additional symptoms were lack of appetite, bloating, frequent cramps in her calves, and problems with her hair. She also mentioned that she suffered from inflammation of the joints and rheumatism.

She therefore was taking the medications prednisolone, pantoprazole, hydroxychloroquine, folic acid and Vigantoletten 2000 IU. She continues to receive a methotrexate injection (MTX) once a week.

Her nutritional anamnesis shows consumption of a lot of empty carbohydrates. She reports eating a lot of fruit during the day and a half a chocolate bar in the evening.

The cortisone intake and regular injections of MTX to reduce RA symptoms are strictly treating symptoms. The RA factors are within normal limits. At the same time the drug pantoprazole is prescribed to minimize the effects on the stomach that are side effects of the other medications. Next to the occurrence of severe pain there are also the reported symptoms of inner restlessness, trouble sleeping, as well as problems with the hair structure.

Our results show a significantly reduced testosterone concentration (fig. 115), which is critical to dihydrotestosterone and thus the hair structure. After only 10 weeks, the targeted intake of the amino acid arginine with 4 g in the morning and 4 g midday shows a verifiably improved hair structure because the natural amino acid indirectly activates HGH (human growth hormone).

A primary measurement of the arginine concentration of 0.49 mg/dl can result in severe problems in the hair structure. Arginine concentrations of 2.1 mg/dl verifiably stabilize the hair structure.

Results of the blood test beforehand

TSH-basal value	0,42 µIU/ml	Calcium	2,39 nmol/l
Ferritin	342,6 ng/ml	Hematocrit	44,3 %
Transferrin	280 mg/dl	Leucocytes	6,89 Tsd/µl
Cholesterol	198 mg/dl	Erythrocytes	4,63 Mio/µl
Uric acid	5,6 mg/d	Thrombocytes	337 Tsd/µl
HDL	70 mg/dl	Hemoglobin	12,9 g/dl
LDL	109 mg/dl	MCV	95,7 fl
Triglycerides	97 mg/dl	MCH	27,9 pg
HbA_{1c}	5,4 % (35,4 mM/M)	MCHC	29,1 g/dl
CRP	< 0,30 mg/dl		
Vitamin D	43,4 nmol/l		
Urea	3,7 mg/dl	Zonulin: 23.30 ng/ml	
Uric acid	3,2 mg/dl	Thyroid antibodies TPO, TRAK: normal	
Creatinine	0,57 mg/dl	Hormone status: beginning menopause	
Sodium	142 nmol/l	Significantly reduced testosterone level 0.06 ng/ml	

Fig. 115: Lab parameters before micronutrient therapy of a 46-year-old woman with severe menopausal symptoms

But the side effects from the previously taken medications are much more severe. The borderline reduced thyroid hormones were not taken into consideration during the previous treatment. But an autoimmune disease of the thyroid can most likely be ruled out since the antibodies TPO and TRAK are within normal range. A TSH value of 0.45 µIU/ml results in a high stress index (fig. 116), a significantly reduced pNN50 (fig. 117) and an elevated vegetative quotient (LF/HF ratio; fig. 118).

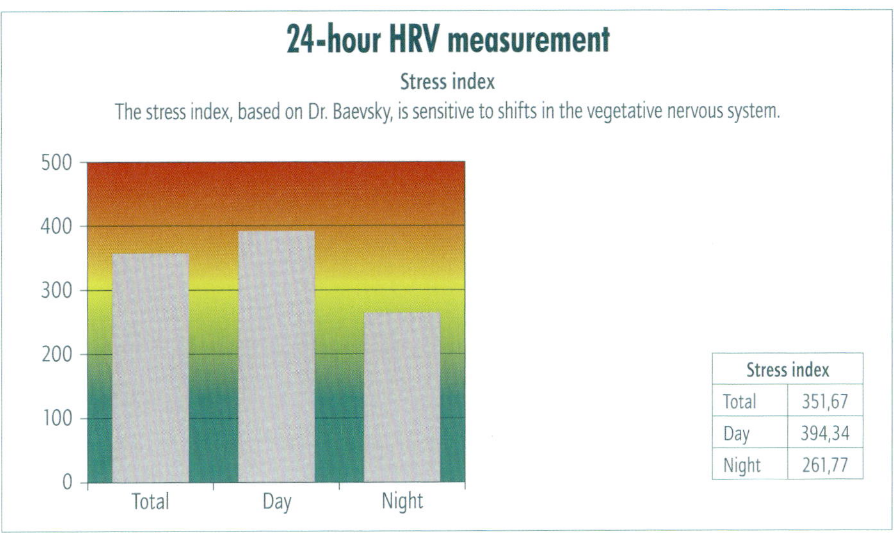

Fig. 116: Stress index before micronutrient therapy in a 46-year-old woman with severe menopausal symptoms

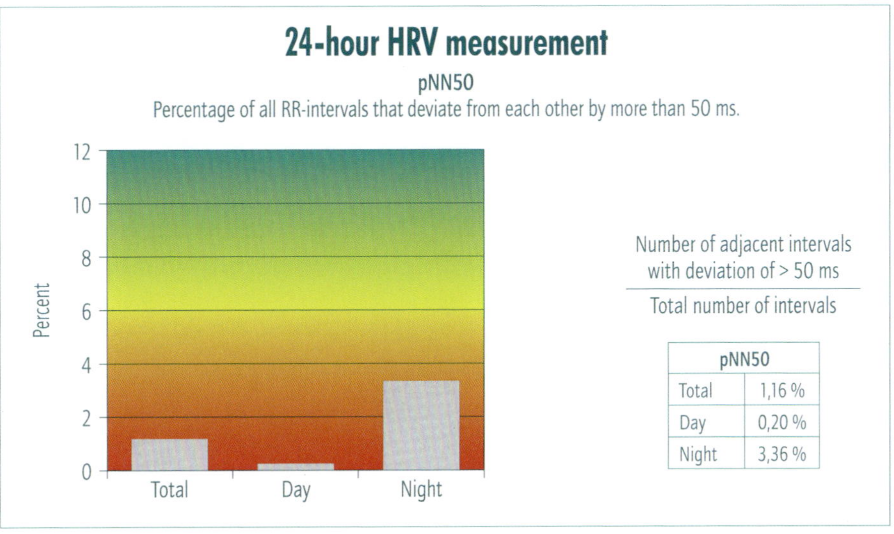

Fig. 117: pNN50 before micronutrient therapy in a 46-year-old woman with severe menopausal symptoms

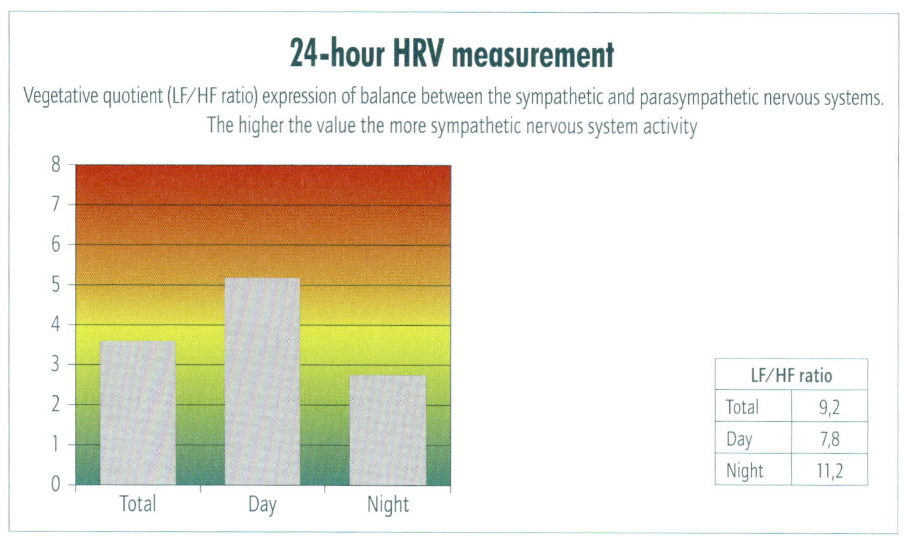

Fig. 118: Vegetative quotient before micronutrient therapy in a 46-year-old woman with severe menopausal symptoms

The primary task of micronutrient therapy is to regulate the thyroid hormones and to normalize the vegetative nervous system. The 46-year-old woman's chronogram of the 24-hour HRV measurement (fig. 119) shows no real opportunity for relaxation day or night.

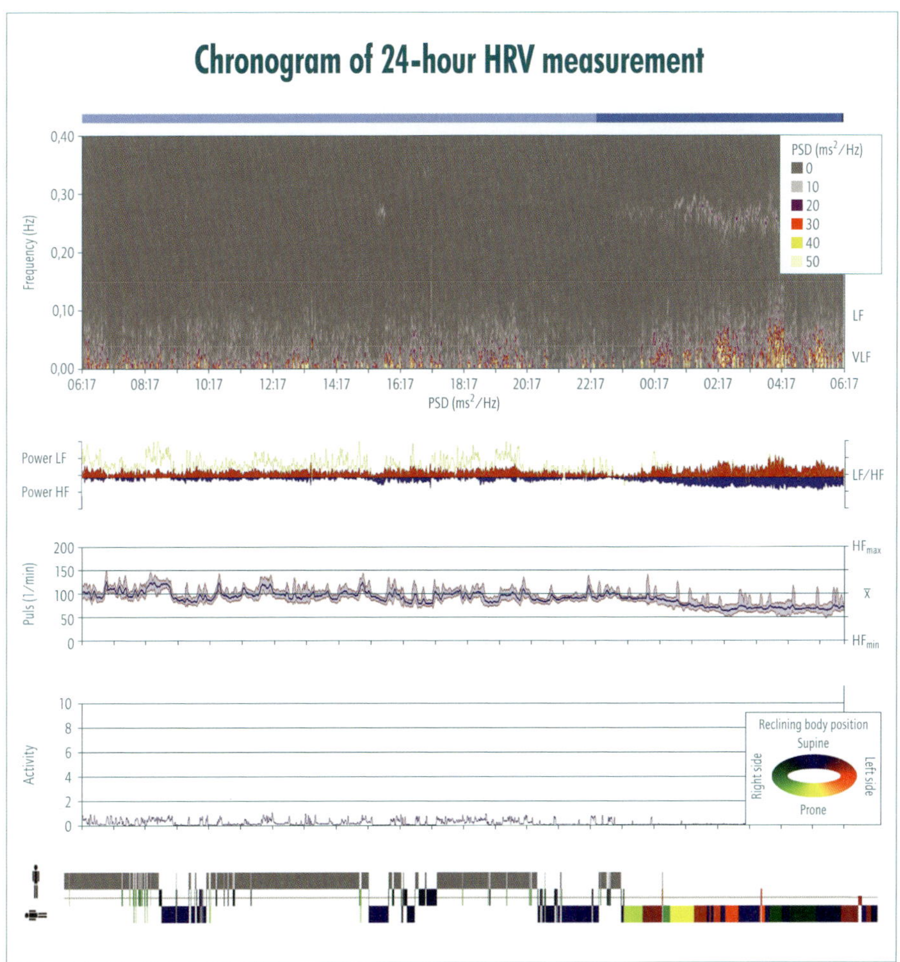

Fig. 119: Chronogram of 24-hour HRV measurement before micronutrient therapy in a 46-year-old woman with severe menopausal symptoms

The functional energy metabolism shows significantly elevated citric, succinic, fumaric, and malic acid levels, an elevated lactate and pyruvate concentration with severe limitations to the activity of certain enzymes (fig. 120).

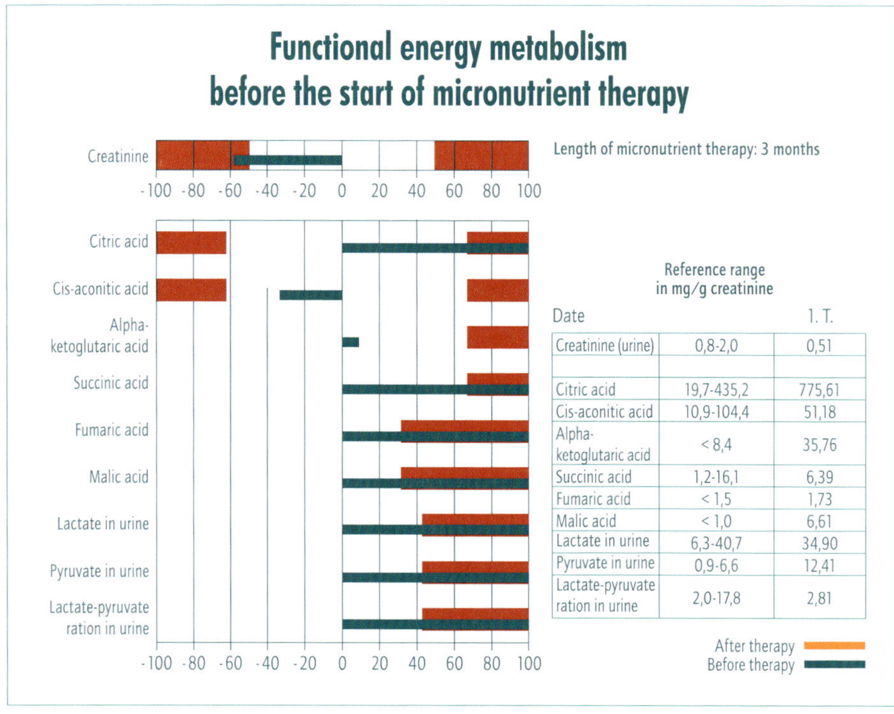

Date	Reference range in mg/g creatinine	1. T.
Creatinine (urine)	0,8-2,0	0,51
Citric acid	19,7-435,2	775,61
Cis-aconitic acid	10,9-104,4	51,18
Alpha-ketoglutaric acid	< 8,4	35,76
Succinic acid	1,2-16,1	6,39
Fumaric acid	< 1,5	1,73
Malic acid	< 1,0	6,61
Lactate in urine	6,3-40,7	34,90
Pyruvate in urine	0,9-6,6	12,41
Lactate-pyruvate ration in urine	2,0-17,8	2,81

Fig. 120: Functional energy metabolism before micronutrient therapy in a 46-year-old woman with severe menopausal symptoms

The woman is given a combination of micronutrients from the HCK modular system in the form of a magistral preparation (fig. 121). Based on many years of experience, an additional intake of the amino acid 5-HTP with 100 mg each, morning, midday, and evening, verifiably helped to significantly reduce the pain symptoms within four weeks.

The simultaneously administered Vitamin B_{12} concentration was 1890 µg, as the HoloTC concentration (active form of Vitamin B_{12}) was considerably reduced. The patient further received magnesium over the course of the day with the intention of normalizing the sympathicotony (activation of the vegetative nervous system) triggered by the borderline reduced TSH basal value.

Micronutrient formulation for a 42-year-old woman with severe menopausal symptoms

Agent	Daily dose	Agent	Daily dose
Vitamins		**Trace elements**	
Vitamin A (retinol)	1 mg	Chrome	300 µg
Vitamin B_1 (Thiamin)	50 mg	Selenium	200 µg
Vitamin B_2 (riboflavin)	50 mg	Zinc	48 mg
Niacin (Vitamin B_3)	50 mg	**Minerals**	
Vitamin B_6 (pyridoxin)	40 mg	Calcium	400 mg
Vitamin B_{12} (Cyanocobalamin)	1.890 µg	Silicon	40 mg
Vitamin C (ascorbic acid)	3.000 mg	**Quasi-vitamins**	
Vitamin D	200 µg	Choline	160 mg
Natural Vitamin E	165,6 mg	Inositol	120 mg
Of that alpha-tocopherol	95,2 mg	PABA	40 mg
Gama-tocopherol	39,1 mg		
Natural carotenoid	8 mg	**Plant extracts**	
Of that alpha-carotin	70 µg	Citrus flavonoids	200 mg
Beta-carotin	1,9 mg	**Fiber prebiotics**	
Lutein	6 mg	Cellulose (HPM)	86 mg
Zeaxanthin	300 µg	Galactomannan	8.442,4 mg
Biotin (Vitamin H)	160 µg	Inulin	59,3 mg
Folic acid (Vitamin B_9)	1.600 µg	Lactate (L-(+)-)	458,3 mg
Pantothenic acid	60 mg	**Additives**	
Amino acids and derivatives		Potato starch	426,6 mg
Glutathione	200 mg	**Building compounds**	
Glutamine	2.000 mg	Methylsulfonylmethane	2.000 mg

Daily intake: 30 ml replenishing phase
Week 1: in the morning with breakfast: 15 ml
After approx. one week also midday: 15 ml
Maintenance phase after 10 weeks: 12 ml in the morning and 12 ml midday

Additionally

- Pea protein without tryptophan, 6/6/4
- Magnesium bisglycinate 100 mg 1/1/2
- 2 Tbs Norsam Omega-3 oil (1 Tbs = 1.700 mg EPA/DHA), 1/1/0
- 5-HTP (hydroxytryptophan) Griffonia 100 mg, 1/1/1
- Ubiquinol 50 mg, 2/1/0 (After 12 weeks only 2 x 50 mg in the morning)

Fig. 121: Micronutrient formulation for a 42-year-old woman with severe menopausal symptoms

After eight weeks the thyroid hormones, particularly the TSH-basal level, rose to 1.9 µIU/ml. Based on the latest scientific studies (Wienecke, 2018; Wienecke, 2020) a "sense of wellbeing" can be achieved in this reference range (➨ chapter 3.1).

Moreover, the targeted intake of hydrolyzed pea protein improves the myofascial system's elasticity.

But here it is particularly important that the amino acid combination does not include L-tryptophan as this can cause a cascade of neurotoxic reactions. The customized micronutrient formulation was developed based on data from the database of comparable patients in this age group with disorders.

After eight weeks of micronutrient therapy the patient's quality of life was significantly improved, and she was also able to keep her composure in stressful situations. The immune system was significantly improved, which was also apparent in the reduced allergy symptoms. Additionally, she had hardly any rheumatoid symptoms in spite of the patient having discontinued prednisolone, the MTX injections, and the gastroprotective pantoprazole.

The timely detection and correction of biochemical structures had a verifiably positive effect on this patient's quality of life. The patient's blood test results compared to those of women in her age group with similar symptoms also show striking deficiencies specifically in the brain activating amino acids phenylalanine and tryptophan (tab. 24 and fig. 122). As previously mentioned, L-tryptophan is administered only as 5-HTP.

Tab. 24: Amino acids with precursor substance for calming and activating neurotransmitters of the brain metabolism

	Laboratory Reference ranges (mg/dl)	Value measured in the patient (mg/dl)	Median value of 450 comparable women in this age structure, symptoms, preexisting conditions, athletic activity
Phenylalanine	0,743 – 2,387	0,690	2,135
Tryptophan	0,715 – 1,829	1,175	1,732

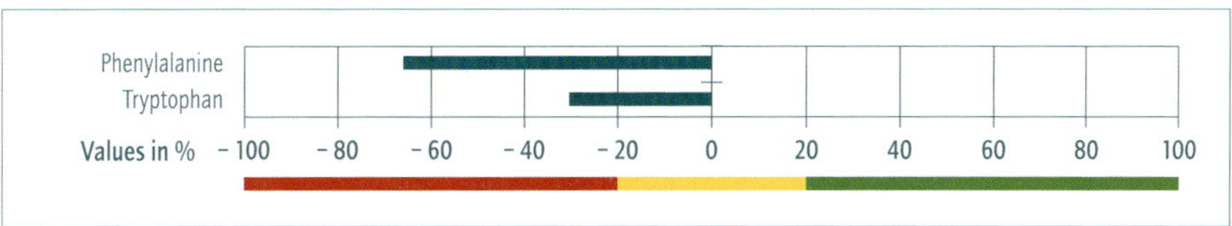

Fig. 122: Deviations of amino acids from the baseline compared to the median value of comparable persons in that age structure, state of health, preexisting conditions, gender

6.3.7 Sleep disorders

"Lack of nightly rest: Germany does not sleep well!"

We see this headline all the time on social media. Nearly every other person in our population has trouble falling asleep or staying asleep. Sleep comes too late, and waking up too early. Millions of people in Germany suffer from insomnia. One of the possible consequences: A significant increase in stress hormones leads to long-term exhaustion and verifiably to dysregulation of the vegetative nervous system all the way to mood fluctuations and many disorders.

But the right dosage of micronutrients will let you sleep!

Using the 24-hour HRV measurement, we measured the stress index, pNN50, and the vegetative quotient (LF/HF ratio) in more than 330 women and men (age structure: 43,2 Jahren ± 7,8) and quantified the relationship between sleep behavior, thyroid hormones, and the cellular level of micronutrient concentrations.

90% of tested individuals show a TSH-basal value of < 1.3 µIU/ml or > 2.5 µIU/ml (tab. 25). Balanced sleep behavior can be seen in individuals whose TSH-basal values lie between 1.6 and 2.2 µIU/ml. A customized micronutrient intake can normalize or harmonize thyroid hormones.

Tab. 25: Thyroid hormones and balance of the vegetative nervous system before and after the customized micronutrient intake in 330 persons with insomnia

	TSH- basal values before N = 135 0,97 ± 0,33	TSH- basal values after N = 135 1,58 ± 0,19	TSH- basal values before N = 195 3,51 ± 0,39	TSH- basal values after N = 195 1,98 ± 0,22
Stress index	455 ± 34,6	153 ± 22,1	479 ± 78,1	170 ± 38,1
pNN50	6,31 ± 2,4	13,5 ± 2,3	5,99 ± 1,7	11,6 ± 2,41
LF/HF-ratio	8,30 ± 3,2	1,91 ± 0,17	7,71 ± 3,6	2,01 ± 0,11

These will result in verifiably improved sleep behavior, which can be verified via the 24-hour HRV measurement. The stress index is significantly reduced both during the day and at night (tab. 25 and fig. 123). The vegetative quotient (LF-HF-ratio; tab. 25 and fig. 125) is significantly decreased while the pNN50 (tab. 25 and fig. 124) as an indicator for the parasympathetic nervous system is significantly increased, thereby documenting a well-balanced vegetative nervous system. This verifiably leads to better night-time sleep behavior.

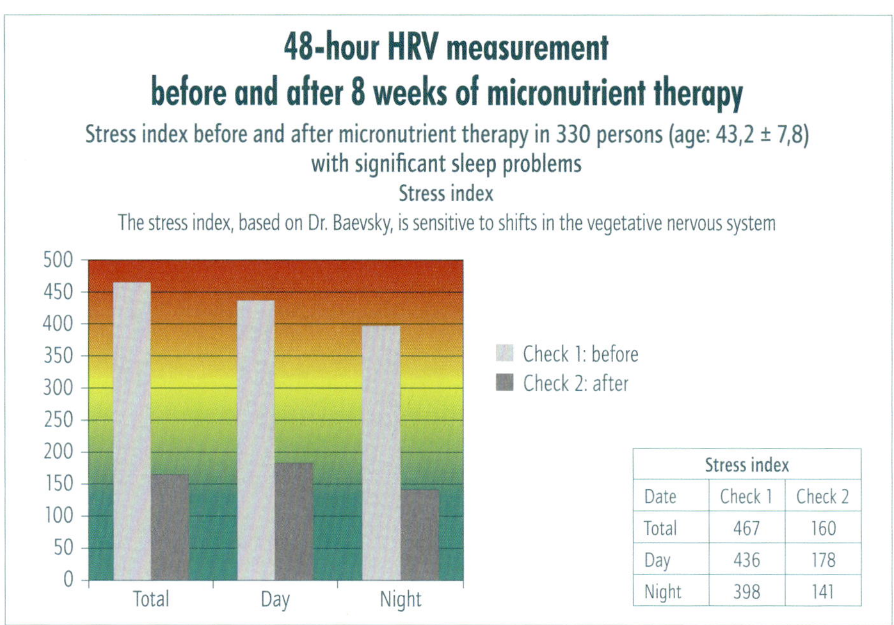

Stress index		
Date	Check 1	Check 2
Total	467	160
Day	436	178
Night	398	141

Fig. 123: Stress index before and after a customized micronutrient intake in 330 persons with insomnia

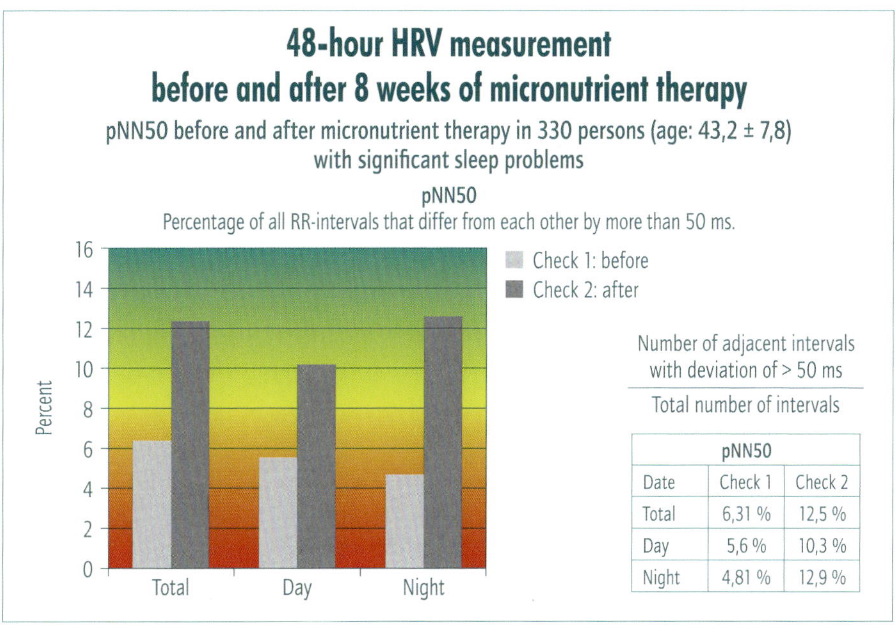

pNN50		
Date	Check 1	Check 2
Total	6,31 %	12,5 %
Day	5,6 %	10,3 %
Night	4,81 %	12,9 %

Fig. 124: pNN50 before and after customized micronutrient intake in 330 persons with insomnia

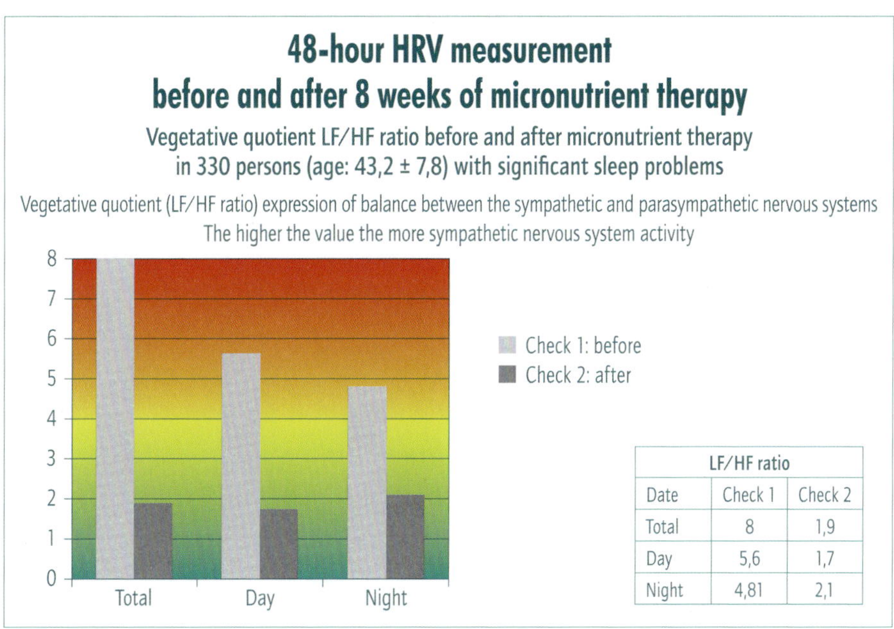

LF/HF ratio		
Date	Check 1	Check 2
Total	8	1,9
Day	5,6	1,7
Night	4,81	2,1

Fig. 125: Vegetative quotient before and after customized micronutrient intake in 330 persons with insomnia

THE POWER OF MICRONUTRIENTS

In addition to the customized micronutrient formulations based on the HCK modular system, an additional intake of magnesium administered throughout the day and the amino acids 5-HTP and ornithine are especially important.

PLEASE NOTE

The amino acid 5-HTP should not be taken while regularly taking antidepressants or antiepileptics. Dosages of micronutrients also depend on blood test results.

Sleep hygiene is essential for people suffering from insomnia. According to the German Society for Sleep Research and Sleep Medicine (https://www.dgsm.de/downloads/dgsm/arbeitsgruppen/ratgeber/neu-Nov2011/Schlafstoerung_A4.pdf) this includes a comfortable bed in a darkened and quiet and rather cool room with fresh air. No stimulants such as coffee or nicotine before going to bed. Alcohol helps to fall asleep faster but causes sleep interruptions and waking early. Relaxing rituals before going to bed and regular wake-up times promote healthy sleep.

PLEASE NOTE

Avoid any "electrosmog" at night. Please, no television in the bedroom and avoid all electronic devices as these cause the verifiable activation of the sympathetic nervous system and prevent a slow-wave sleep phase. And please, don't keep your cellphone in the bedroom.

6.3.8 Competitive and elite sports

Achieve optimal mental and physical performance capacity with the right dosage of lacking micronutrients!

In elite athletes, too, the growing mental and physical demands increasingly lead to exhaustion, major performance and mood fluctuations, all the way to inexplicable injuries. In recent years, more than 14,450 performance and elite athletes were tested by SALUTO (tab. 26), giving us the ability to provide an overview of the athletes' energy balance before and after micronutrient supplementation (fig. 126).

Tab. 26: Anthropometric data of 14,450 tested performance and elite athletes between 1994 and 2019

	Age (years)	Height (cm)	Weight (kg)	BMI (kg/m^2)
Total (N = 14.450)	26,3 ± 9,9	18,0 ± 4,5	71,45 ± 22,05	22,05
Female (N = 6.941)	24,3 ± 7,3	175 ± 4,3	68,1 ± 4,2	22,3
Male (N = 7.509)	28,2 ±8,3	185 ± 4,8	74,3 ± 4,6	21,9

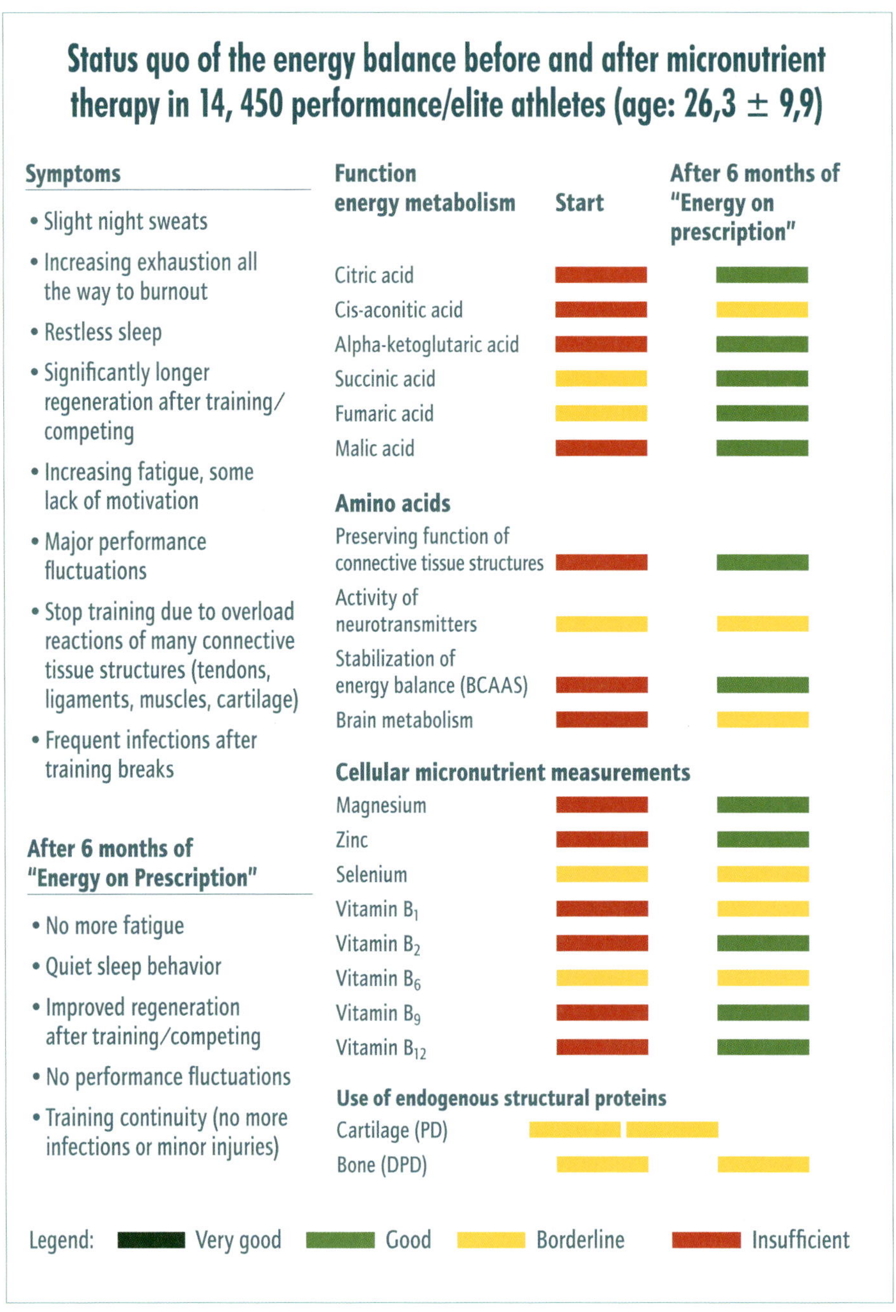

Fig. 126: Status quo of energy balance before and after customized micronutrient intake in 14,450 performance/elite athletes

Next to many psychological and mental training methods for optimizing balance of the vegetative system there are scientifically valid measuring instruments that make it possible to ascertain detailed information about many vegetative nervous system parameters. The 48-hour HRV analysis allows us to evaluate the performance and recovery state, the effect of training, regeneration ability as well as the associated sleep quality (fig. 127, 128, and 129.)

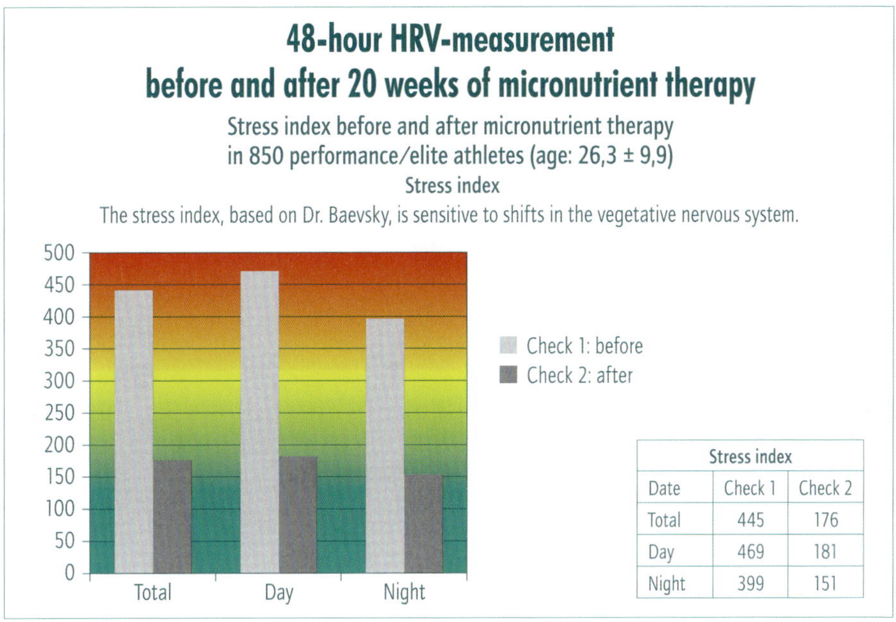

Stress index		
Date	Check 1	Check 2
Total	445	176
Day	469	181
Night	399	151

Fig. 127: Stress index before and after customized micronutrient intake in 850 performance-/elite athletes

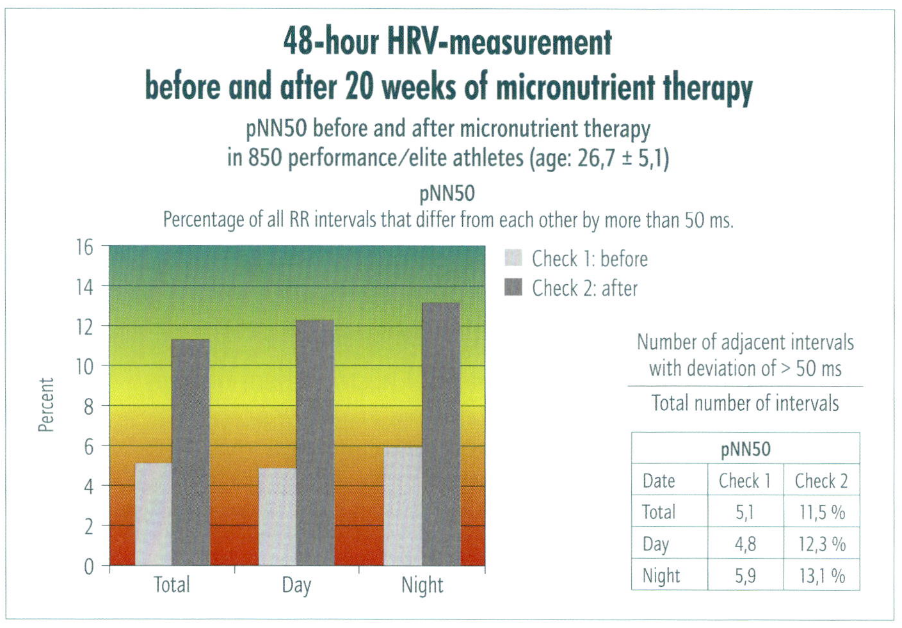

pNN50		
Date	Check 1	Check 2
Total	5,1	11,5 %
Day	4,8	12,3 %
Night	5,9	13,1 %

Fig. 128: pNN50 before and after customized micronutrient intake in 850 performance/elite athletes

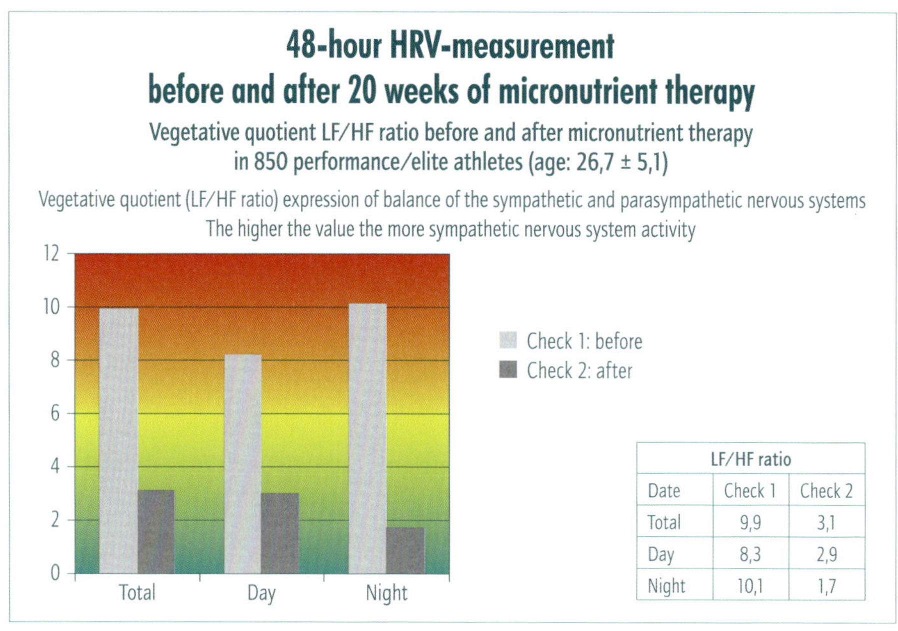

LF/HF ratio		
Date	Check 1	Check 2
Total	9,9	3,1
Day	8,3	2,9
Night	10,1	1,7

Fig. 129: Vegetative quotient before and after the customized micronutrient intake in 850 performance/elite athletes

Furthermore, psychophysiological disorders can be detected at an early stage. To date the literature does not include any long-term analyses about the extent to which optimized nutritional behavior impacts many of the 48-hour HRV measurement parameters over longer periods of time.

6.3.8.1 The effect of an optimal diet and customized micronutrient supply on competitive and elite athletes

An evidence-based, retrospective study focused on the impact of an optimized diet and a customized micronutrient intake on parameters of the vegetative nervous system in performance/elite athletes with the aid of the 48-hour HRV measurement and the pyridinium crosslinks. Both groups capture the athletes that optimized their nutritional behavior over the course of a two-year testing period (anthropometric data: tab. 27 and 28), during the first year without a targeted micronutrient formulation and during the second year with a customized micronutrient formulation (fig. 130).

Tab. 27: Anthropometric data from 158 marathon runners

	Without supplementation N = 158	With supplementation N = 158
Age	27,12 ± 3,13	27,49 ± 3,08
Height	179,37 ± 6,33	179,37 ± 6,33
Weight	78,64 ± 4,75	79,28 ± 5,28

Tab. 28: Anthropometric data from 144 pro soccer players

	Without supplementation N = 144	With supplementation N = 144
Age	27,23 ± 2,94	28,23 ± 2,94
Height	179,67 ± 6,25	179,67 ± 6,25
Weight	79,11 ± 4,72	78,83 ± 4,62

THE POWER OF MICRONUTRIENTS

The "carbohydrate myth" is in the past

The tested athletes had a very carbohydrate-conscious diet. The motto "more is better" is now scientifically obsolete and disproven. Especially during regeneration phases, the body requires high-quality structural proteins and fats (e.g. Omega-3 fatty acids that have a verifiable anti-inflammatory effect). The beloved pasta party the night before a competition/game is in the past as well as the perpetual intake of carbohydrates from sources like noodles, potatoes, and rice, the frequent consumption of which can verifiably lead to increased inflammation in athletes.

Lectins from wheat and rye verifiably cause:

- Increased inflammation;
- An attack on joint structures, as lectins bind acetylglucosamine and the preservation of function of stressed structures is thus no longer possible.
- Phytic acid decreases the vitamin/mineral status because minerals are bound. **Aside:** Athletes with iron problems should therefore consume sourdough from oats or spelt.
- Gluten in grain attacks the intestinal mucosa (gluten-accompanying substances: ATPA, adenosine triphosphate amylase; "Today's wheat has a more aggressive effect on the gut").

The athletes take into account a balance between carbs with a high glycemic index (quick and high increase in blood sugar, high insulin load, quick absorption into muscle cells) and carbs with a low glycemic index (slow and low increase in blood sugar, steady supply of glucose to muscle cells).

The athletes in this study ate according to the principle: carbohydrates make sense during the biggest physical strain. But they should be used with the utmost caution during breaks and phases with a lower training load. A combination of high glycemic and low glycemic carbohydrates is used during competitions.

Eat fruit, but the right way!

The athletes stopped eating fruit in the evening. Eating apples, bananas, pineapple, and pears in the evening causes a major insulin response and thus verifiably poor sleep- and regeneration behavior. The athletes ate primarily different types of berries (huckleberries, blueberries, strawberries, and blackberries) mostly during the morning hours. These contain less fructose. Fruit should ideally be consumed at mealtime and not, as used to be the case, throughout the day.

Fit with fats?

The athletes participating in the study preferred high-quality fats and largely avoided margarine, cookies, crackers, processed foods, chips, and peanut puffs. The results from the athletes' nutritional analysis show a 2:1 ratio of Omega-6 to Omega-3 fatty acids. That is near perfect for athletes.

Further results of the athletes' nutritional analysis show that a diet with more fat and simultaneously fewer carbohydrates can have a performance-enhancing effect in sports. Higher fat oxidation during physical exertion conserves glycogen reserves and thereby ensures sufficient energy reserves, for instance for sprint loads.

The nutritional behavior of the individual athletes was adapted to their respective training and competition phases; carbohydrate modification during intensive training/competition phases, high in protein and fat optimized.

The nutritional analyses are done with the nutritional value calculator software *Opti-Diet* by GOE. The analysis is based on a 7-day dietary record. It was completed three times by the runners (N = 158) as part of their preparation for a marathon. The pro soccer players (N = 144) completed the test five times at different testing dates over the course of the season.

In summary, we present several nutritional aspects the athletes integrated into their everyday training that correspond to the key principles of Dr. Mosetter's Glycoplan.

- Fluid intake 40 ml/kg bodyweight plus 1 l/hour of training.
- Mineral water with a high hydrogen carbon content (> 1500 mg).
- Carbohydrate-conscious (reduced intake of rice, potatoes, pasta, etc.) only during intensive training/competition phases.
- High-quality fats and protein intake while simultaneously consuming fewer carbs, e.g. during regeneration phases.
- Regular intake of galactose/ribose (insulin-independent glucose) before and after every training unit to optimize the energy metabolism.
- Lots of vegetables and salad every day.
- Preferred fruits are berries (huckleberries, blackberries, blueberries, strawberries) during morning hours because they have a low glycemic index.
- High-protein and low-carb diet on normal training days and in the evenings (e.g. omelets with vegetables).

The previously described blood and urine analyses are critical for the creation of customized micronutrient formulations. Tab. 29 and fig. 130 show the correct reference ranges for the formulations.

Tab. 29: Dosages of 30-70 g of high-quality amino acid blends

Amino acids	Specifications for 30-70 g intake
L-alanine	2,4-5,6 g
L-arginine	2,8-6,9 g
L-aspartic acid	1,7-4,0 g
L-glutamic acid	4,1-9,5 g
L-glycine	6,0-14,0 g
L-histidine	0,4-0,92 g
L-isoleucine	1,4-3,3 g
L-leucine	1,8-4,2 g
L-lysine	1,1-2,5 g
L-methionine	0,2-0,4 g
L-phenylalanine	0,6-1,4 g
L-proline	3,84-9,0 g

Amino acids	Specifications for 30-70 g intake
L-serin	0,5-1,2 g
L-threonine	0,5-1,3 g
L-tyrosine	0,6-0,7 g
L-valine	1,9-4,4 g

Reference ranges of micronutrient concentrations by individual micronutrient formulations based on the HCK modular system

Agent	Daily dosw
Vitamins	
Vitamin A (retinol)	0,8-1,5 mg (1.000-3.600 i.E.)
Vitamin B_1 (thiamine)	10-70 mg
Vitamin B_2 (riboflavin)	10-70 mg
Vitamin B_6 (pyridoxin)	20-80 mg
Vitamin B_{12} (cyanocobalamin)	30-1.500 µg
Vitamin C (ascorbic acid)	200-4.000 mg
Vitamin D_3	50-200 µg
Natural Vitamin E	150-400 mg
Of that alpha-tocopherol	85-340 mg
Gamma-tocopherol	13-53 mg
Natural carotenoid	8 mg
Of that alpha-carotin	70 µg
Beta-carotin	1,9 mg
Lutein	6 mg
Zeaxanthin	15,0 µg
Biotin (Vitamin H)	45-150 µg
Folic acid (Vitamin B_9)	400-1.600 µg
Niacin (Vitamin B_3)	10-60 mg
Pantothenic acid	20-80 mg
Additional amino acids in special cases	
Arginine	1.000-3.000 mg
Glutamine	500-3.000 mg
5-HTP	100-400 mg
Glutathione	100-300 mg
MSM	1.000-3.000 mg

Agent	Daily dose
Trace elements	
Chrome	100-500 µg
Manganese	5-20 mg
Molybdenum	50-200 µg
Selenium	50-200 µg
Zinc	12-48 mg
Minerals	
Calcium	200-800 mg
Potassium	200-400 mg
Magnesium	200-600 mg
Silicon	20-60 mg
Quasi vitamins	
Choline	80-240 mg
Coenzym Q_{10}	30-200 mg
Inositol	60-180 mg
L-carnitine	250-1.000 mg
PABA	mg
Plant extracts	
Green tea extract	mg
Citrus flavonoids	mg
Red wine extract	mg
Fiber prebiotics	
Guar gum	mg
HPM cellulose	mg
Inulin	mg

Fig. 130: Reference ranges of micronutrient concentrations by individual micronutrient formulation based on the HCK modular system

Below is a presentation of the impact of customized micronutrient formulations on the cellular micronutrient concentrations and the parameters of the 48-hour HRV measurement (stress index, pNN50, vegetative quotient, LF/HF-ratio) that have resulted in much improved long-term stress tolerance with respect to training and competition loads, first using the example of the 158 marathon runners.

6.3.8.2 Marathon runners during preparation

Runners at a performance level between 2:30 and 2:50 h marathon running time who have demonstrated balanced nutritional behavior are captured evidence-based and retrospectively during preparation for a marathon (N = 158).

Detailed findings about the progression of the different vegetative nervous system parameters can be ascertained at different points during the preparation phase to the marathon. The 48-hour HRV analysis always takes place from Tuesday to Thursday morning: before the start of preparation, after 8, 16, and 24 weeks.

In addition, the runners are asked to fill out a special questionnaire for individualized evaluation. The test for homogeneity shows no statistically significant differences for the individually measured parameters at the beginning of preparation. After eight weeks of intensive training, significant differences in the listed parameters can verifiably be detected between the first measuring time point (MTP) and the second MTP (tab. 30).

Tab. 30: Impact of optimized nutritional behavior and a targeted micronutrient intake on the vegetative nervous system's parameters with the aid of the 48-hour HRV measurement in 158 runners during marathon preparation (performance level: 2:30-2:50 h)

Parameter		Start	After 8 weeks	After 16 weeks	After 24 weeks	Percentage change p $p < 0{,}001$ highly significant***	Test for homogeneity between the groups
	Micro-nutrient intake	1. MTP	2. MTP	3. MTP	4. MTP	1. MTP → 4. MTP	1. MTP → 4. MTP
Sress index	without	363,7 ± 59,8	348,6 ± 62,2	354,1 ± 63,2	378,9 ± 63,9	4,16***	p < 0,001
	with	366,9 ± 56,5	186,5 ± 43,9	187,3 ± 50,7	178,7 ± 48,1	- 51,15***	p < 0,001
pNN50 (%)	without	7,03 ± 0,60	7,36 ± 0,16	7,35 ± 1.02	7,36 ± 0,99	3,52***	p < 0,001
	with	7,34 ± 0,66	12,43 ± 0,73	13,08 ± 0,99	13,37 ± 0,83	86,79***	p < 0,001
HF (ms^2)	without	88,5 ± 8,1	101,5 ± 8,9	107,2 ± 15,2	118,5 ± 16,6	33,45***	p < 0,001
	with	88.6 ± 7.0	503.5 ± 6.8	515.5 ± 11.1	505.5 ± 17.5	474.69***	p < 0.001
LF (ms^2)	without	655,5 ± 73,7	653,0 ± 80,5	673,0 ± 84	706,5 ± 84,5	497,54***	p < 0,001
	with	642,5 ± 83,7	852,5 ± 73	848,7 ± 69	854,4 ± 74,7	86,79***	p < 0,001
LF/HF ratio	without	7,3 ± 0,63	6,50 ± 1,02	6,55 ± 1,4	6,09 ± 1,1	- 16,86***	p < 0,001
	with	7,6 ± 1,2	1,69 ± 0,16	1,88 ± 0,28	1,70 ± 0,17	- 76,76***	p < 0,001

The stress index (fig. 131) of 363.7 ± 59.8 at the first MTP before the start of the preparation phase in the runners without a targeted micronutrient formulation did not statistically change after 8 weeks at 348.6 ± 62.22, while the runners with a targeted micronutrient formulation showed a highly significant ($p < 0.001$) reduction from 366.9 ± 56.5 to 186.5 ± 43.9.

THE POWER OF MICRONUTRIENTS

From the second MTP to the fourth MTP the runners without a micronutrient formulation show a statistically higher stress index than the runners who received targeted micronutrient formulations during the second season. No additional statistical differences in the stress index are discernable from the second to the fourth MTP. A low stress index verifiably indicates an improved stress tolerance. This is also documented by the pNN50, which is considered a stable indicator for parasympathetic nervous system activity (fig. 132).

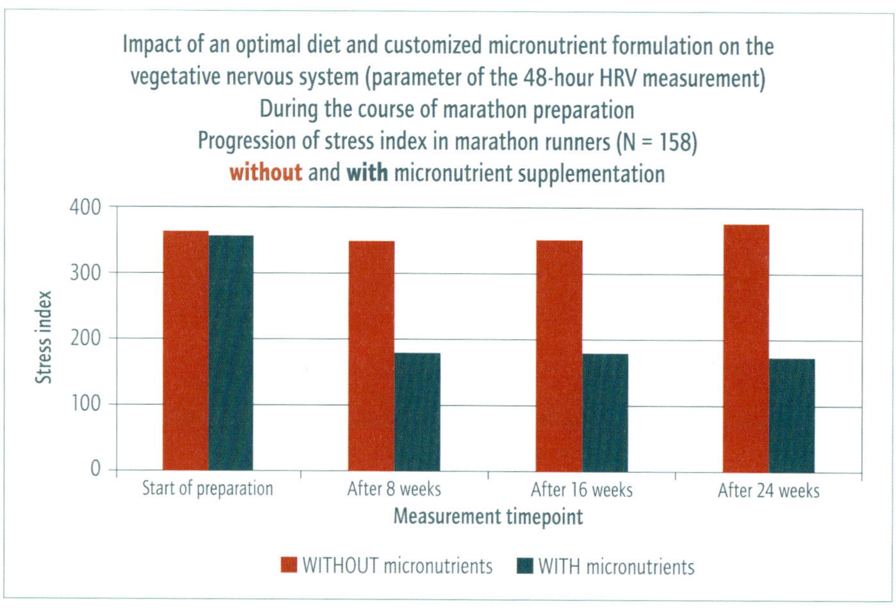

Fig. 131: Stress index before and after customized micronutrient intake in 158 runners during marathon preparation (performance level: 2:30-2:50 h)

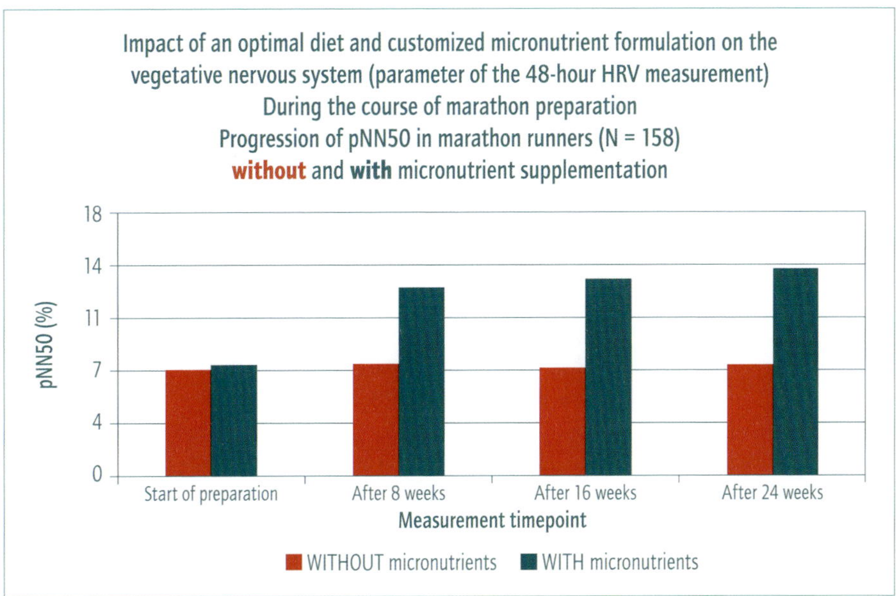

Fig. 132: pNN50 before and after customized micronutrient intake in 158 runners during marathon preparation (performance level: 2:30-2:50 h)

The first MTP with a pNN50 of 7.03 ± 0.60 did not change after the second MTP after eight weeks of an intensive preparation phase with 7.36 ± 0.16 in the runners without a micronutrient formulation, and it shows a disbalance of the vegetative nervous system towards strong activation of the sympathetic nervous system (fig. 132).

85.3% of runners also consistently describe increased inner restlessness combined with very poor sleep behavior, mental and physical overload. The pNN50 changes insignificantly from the second to the fourth MTP. During the second-year preparation phase with targeted micronutrient formulation, the runners show a highly significant increase ($p < 0,001$) from the first MTP with a pNN50 of 7.34 ± 0.66 to a pNN50 of 12.43 ± 0.73 after eight weeks of an intensive preparation phase.

The results clearly reflect the runners' subjective descriptions on the accompanying report sheet. They describe a balanced vegetative nervous system with a good mental and physical performance capacity. The nightly sleep phases also show verifiably reduced activation of the sympathetic nervous system.

The pNN50 also changes insignificantly at the subsequent MTPs after 16 and 24 weeks and shows a well-balanced vegetative nervous system.

This is also confirmed by the LF/HF ratio (fig. 133) as the vegetative quotient that documents the balance between the sympathetic and parasympathetic nervous systems.

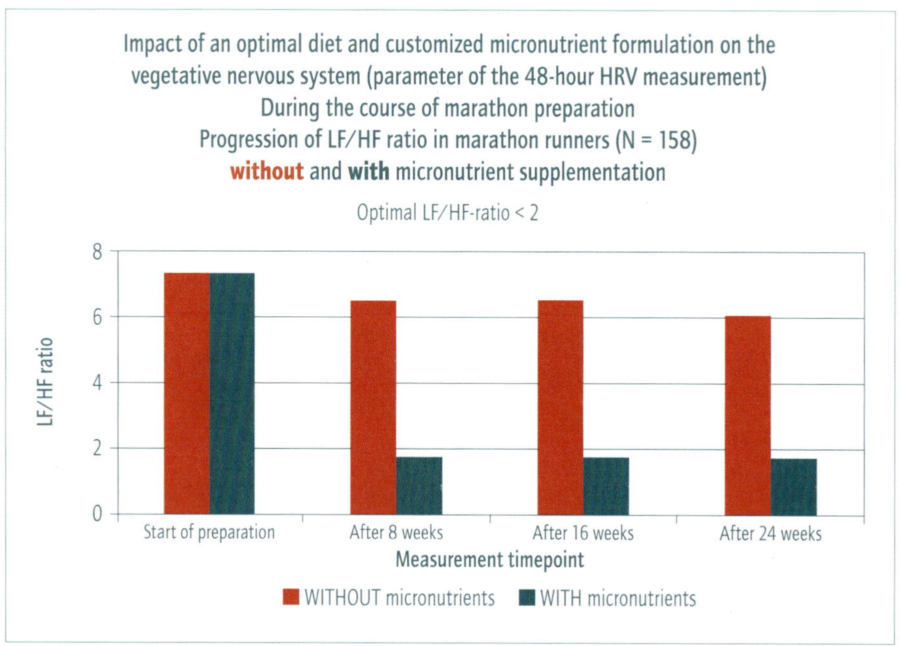

Fig. 133: Vegetative quotient before and after customized micronutrient intake in 158 runners during marathon preparation (performance level: 2:30-2:50 h)

The marathon runners with micronutrient supplementation showed significant improvement of their intracellularly measured micronutrient concentrations (tab. 31 and fig. 134).

Tab. 31: Cellular micronutrient concentrations (M ± SD) of 158 runners during a marathon preparation phase with and without targeted micronutrient supplementation. Percentage changes between the first and fourth MTP as well as key figures from the comparison test between the groups at the fourth MTP

Parameter	Start of preparation (1. MTP)		After 8 weeks of training (2. MTP)		After 16 weeks (3. MTP)		After 24 weeks (4. MTP)		Percentage change (1.MTP → 4. MTP) Highly significant*** p < 0,001	
	without	with	without	with	without	with	without	with	without	with
Mg (mg/l ery.)	42,30 ± 3,57	44,50 ± 3,75	39,40 ± 3,57	48,90 ± 3,12	36,6 ± 3,68	53,61 ± 3,03	33,60 ± 3,62	54,70 ± 3,62	- 10,18***	22,93***
Zn (mg/l ery.)	11,3 ± 0,21	11,0 ± 0,53	9,60 ± 0,76	13,40 ± 0,40	9,30 ± 0,59	13,8 ± 0,71	8,80 ± 0,93	14,3 ± 0,64	- 21,7***	29,22***
Se (µg/l ery.)	81,8 ± 7,86	82,9 ± 8,13	77,5 ± 6,01	119,15 ± 7,63	69,8 ± 7,13	132,53 ± 6,71	56,8 ± 6,78	138,8 ± 8,01	- 26,91***	68.81***
Vit. B_1 (µg/l ery.)	61,62 ± 4,67	62,0 ± 4,11	56,58 ± 4,26	84,80 ± 4,48	52,6 ± 5,28	90,4 ± 4,42	48,66 ± 5,28	93,8 ± 4,01	- 21,03***	54.00***
Vit. B_2 (µg/l ery.)	173,20 ± 23,2	170,62 ± 20,95	159,8 ± 15,81	377,70 ± 18,7	119,5 ± 22,5	430,3 ± 18,0	76,9 ± 24,15	456,50 ± 15,36	- 45,57***	168,81***
Vit. B_9 µg/l ery.	565,0 ± 101	632,9 ± 99,8	533,1 ± 80,75	1.024,18 ± 87,65	506,56 ± 70,36	1.067,27 ±79,3	394,02 ± 90,07	1.092,3 ± 103,6	- 34,34***	73,75***
HoloTC (pmol/l)	78,46 ± 13,42	67,90 ± 8,96	44,44 ± 11,4	131,3 ± 9,81	26,34 ± 10,91	146,45 ± 14,57	23,09 ± 10,41	156,40 ± 14,25	- 71,00***	129,88***

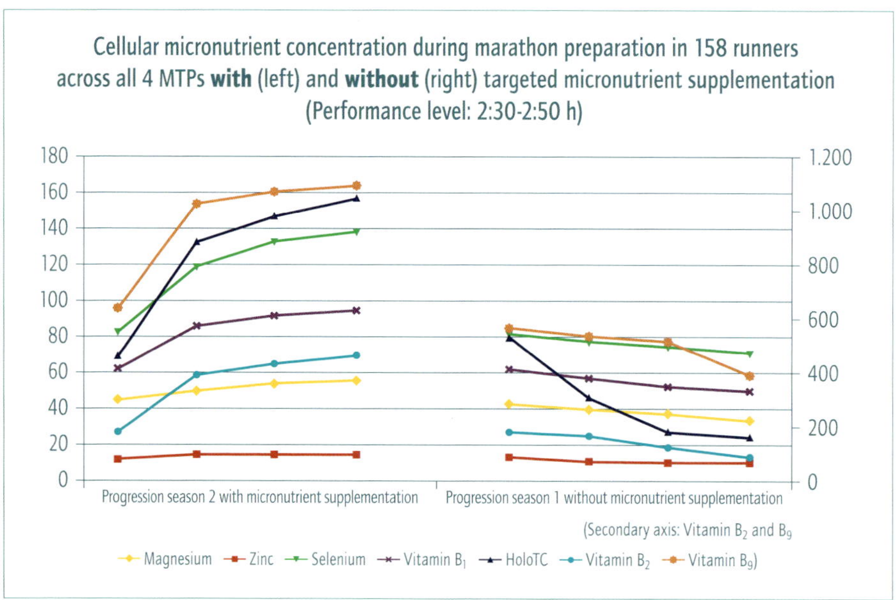

Fig. 134: Micronutrient concentration with and without customized micronutrient intake in 158 runners during marathon preparation (performance level: 2:30-2:50 h)

Comparable results could be seen with respect to the supply of amino acids (tab. 32 and fig. 135).

Tab. 32: Amino acid concentration (M ± SD) of 158 runners during the marathon preparation phase with and without targeted micronutrient supplementation. Percentage changes between the first and fourth MTP as well as key figures from the comparison test between the groups at the fourth MTP

Amino acid (mg/dl)	Start of preparation (1. MTP)		After 8 weeks of training (2. MTP)		After 16 weeks of training (3. MTP)		After 24 weeks of training (4. MTP)		Percentage change (1. MTP → 4. MTP) p < 0,001 Highly significant***		Comparison test between groups 1. → 4. MTP
	without	with	without	with	without	with	without	with	without	with	
Arginine	0,86 ± 0,22	0,83 ± 0,11	0,66 ± 0,17	2,18 ± 0,34	0,50 ± 0,15	2,43 ± 0,41	0,40 ± 0,12	2,76 ± 0,62	-52,80***	231,73***	p < 0,001
Proline	3,09 ± 0,64	3,03 ± 0,89	2,74 ± 0,51	4,61 ± 0,96	2,53 ± 0,55	4,91 ± 1,12	2,38 ± 0,45	5,14 ± 1,24	-22,75***	69,64***	p < 0,001
Methionine	0,63 ± 0,13	0,57 ± 0,23	0,48 ± 018	0,91 ± 0,19	0,39 ± 0,23	1,12 ± 0,32	0,15 ± 0,14	1,20 ± 0,41	-76,75***	110,00***	p < 0,001
Glycine	1,99 ± 0,43	1,96 ± 0,22	1,61 ± 0,31	2,91 ± 0,49	1,39 ± 0,24	3,24 ± 0,74	1,27 ± 0,28	3,51 ± 1,01	-36,34***	79,08***	p < 0,001
Glutamine	6,83 ± 0,97	7,01 ± 0,71	5,85 ± 0,88	9,18 ± 0,79	5,44 ± 0,87	10,59 ± 0,80	5,19 ± 0,68	11,05 ± 0,79	-23,93***	57,74***	p < 0,001
Glutamic acid	3,94 ± 0,47	3,84 ± 0,36	3,04 ± 0,61	6,16 ± 1,47	2,28 ± 0,54	7,50 ±1,68	1,82 ± 0,44	9,14 ± 2,57	-53,71***	137,97***	p < 0,001
Leucine	2,31 ± 0,38	2,26 ± 0,51	1,97 ± 0,42	3,67 ± 0,67	1,77 ± 0,52	3,98 ± 0,89	1,63 ± 0,55	4,12 ± 1,36	-29,57***	82,30***	p < 0,001
Isoleucine	1,35 ± 0,28	1,16 ± 0,45	1,12 ± 0,33	2,49 ± 0,65	0,94 ± 0,17	2,73 ± 0,96	0,81 ± 0,21	2,91 ± 1,27	-40,25***	150,86***	p < 0,001
Valine	3,22 ± 0,39	3,30 ± 0,59	2,98 ± 0,51	3,82 ± 0,69	2,85 ± 0,55	4,09 ± 0,39	2,73 ± 0,32	4,21 ± 1,12	-15,16***	27,58***	p < 0,001

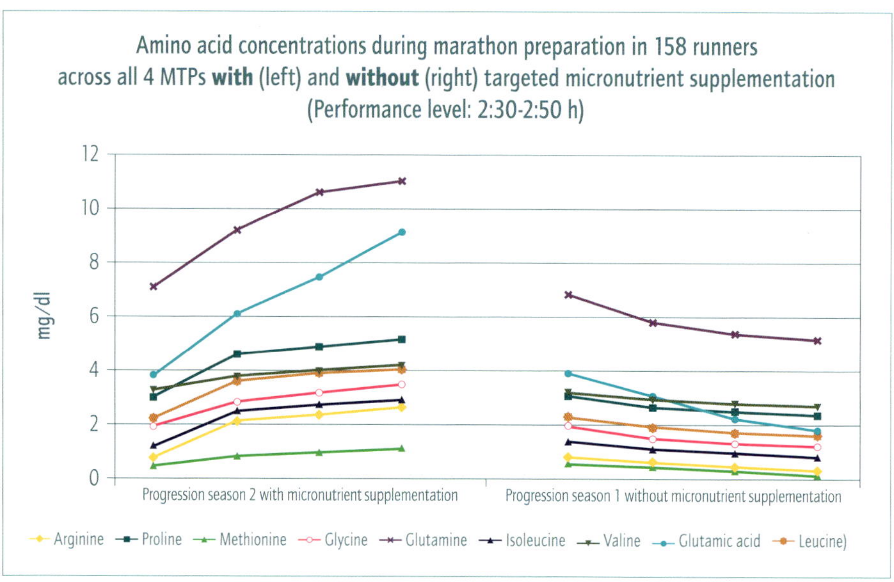

Fig. 135: Amino acid concentrations with and without customized micronutrient intake in 158 runners during marathon preparation (performance level: 2:30-2:50 h)

The individual surveys the runners (N = 158) completed a total of four times show that 87.3% of runners with a targeted micronutrient intake do not show any overload reactions of loaded connective tissue structures (tendons, ligaments, muscles, etc.). Only 2.7% of runners showed the occasional slight muscle discomfort which, however, did not result in missed training.

These runners also do not report any major loss of mental and physical performance before and after the marathon. The highly significant positive changes of the 24-hour HRV-measurement in the sum score and the impact of customized micronutrient formulations can be seen in tab. 31 and tab. 32 as well as fig. 134, 135 and 136.

During the first year of the study, 84.5% of runners from the group of 158 runners without a micronutrient formulation show repeated and increasing infections, overload reactions of many loaded connected tissue structures, that led to missed training of up to four weeks. Additional symptoms include frequent irritation at the patella, the Achilles tendon, and various problems in the knee joints.

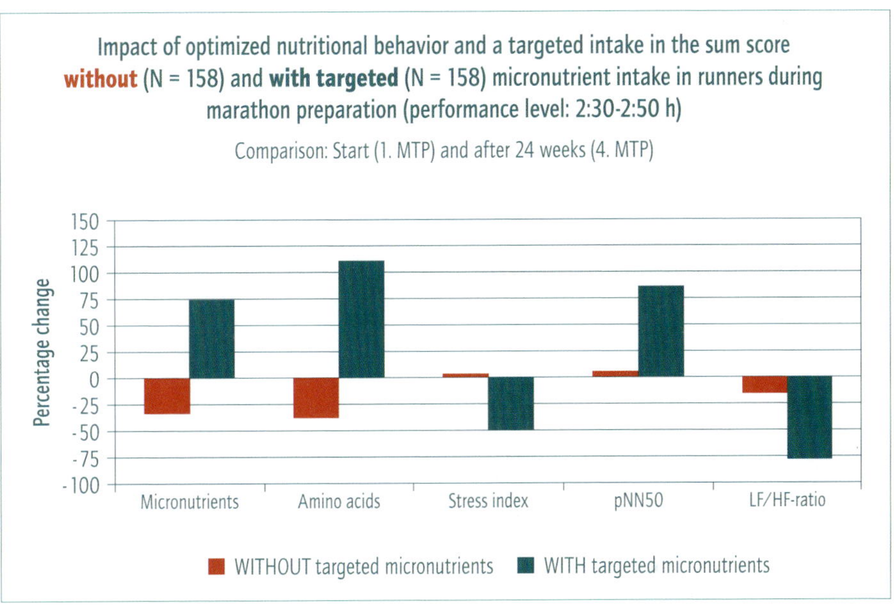

Fig. 136: Percentage change of micronutrients and amino acids as well as different HRV parameters during marathon preparation

CONCLUSION

In the marathon runners, optimized nutritional behavior and a customized micronutrient formulation for each athlete leads to a significant positive change in many parameters of the 48-hour HRV measurement at different MTPs and shows the positive impact on the balance of the vegetative nervous system. A good level of amino acids and cellular micronutrients results in a highly significant reduction in the stress index ($p < 0,001$) raises the pNN50 ($p < 0.001$), and reduces the vegetative quotient (LF/HF-ratio; $p < 0.001$).

These scientifically valid parameters confirm the athletes' subjective assertions in the completed surveys. 91.3% of athletes with a targeted micronutrient intake describe during the entire testing period, or rather the individual MTPs, a considerably improved stress tolerance, much better regeneration after stresses and strains, and show consistent mental and physical performance readiness.

86.7% of athletes who demonstrated optimized nutritional behavior, but did not receive a targeted micronutrient formulation, describe increasing mental/physical exhaustion, struggle to unwind after intensive training/competition loads, and show significant fluctuations in their performance capacity.

Next to many psychological and mental training methods for optimizing the balance of the vegetative nervous system, the many parameters of the 48-hour HRV measurement show that it is possible to evaluate the state of performance and recovery, the effect of training, the regeneration ability as well as the associated sleep quality.

For instance, the literature lists the L-tryptophan amino acid requirement as between 3-6 mg/kg bodyweight. Our findings in the retrospective study of competitive athletes in recent years show a requirement of up to 7 mg/kg bodyweight. The 91.3% of tested athletes with an adequate tryptophan concentration describe a considerably more positive mood not only at the individual MTPs, but especially over the course of the entire testing period. The athletes were given 5-HTP (hydroxytryptophan) in cases of chronically inflammatory processes with increased IDO (indolamine-2, 3-dioxygenase) as the main regulator of the tryptophan/kynurenine/serotonin balance.

Customized micronutrient therapy in athletes with balanced nutritional behavior helps to prevent physiological maladaptation in the sense of "non-functional overreach" in the long-term, and verifiably reduces an increase of pyridinium crosslinks as an indicator for the use of endogenous structural proteins (fig. 137 and 138).

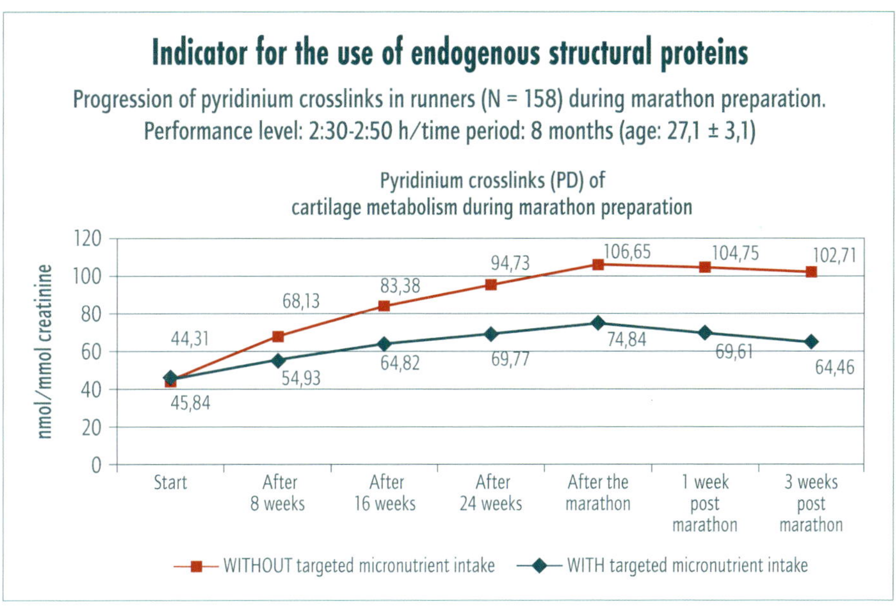

Fig. 137: Progression of pyridinium crosslinks (PD) with and without customized micronutrient intake in 158 runners during marathon preparation (performance level: 2:30-2:50 h)

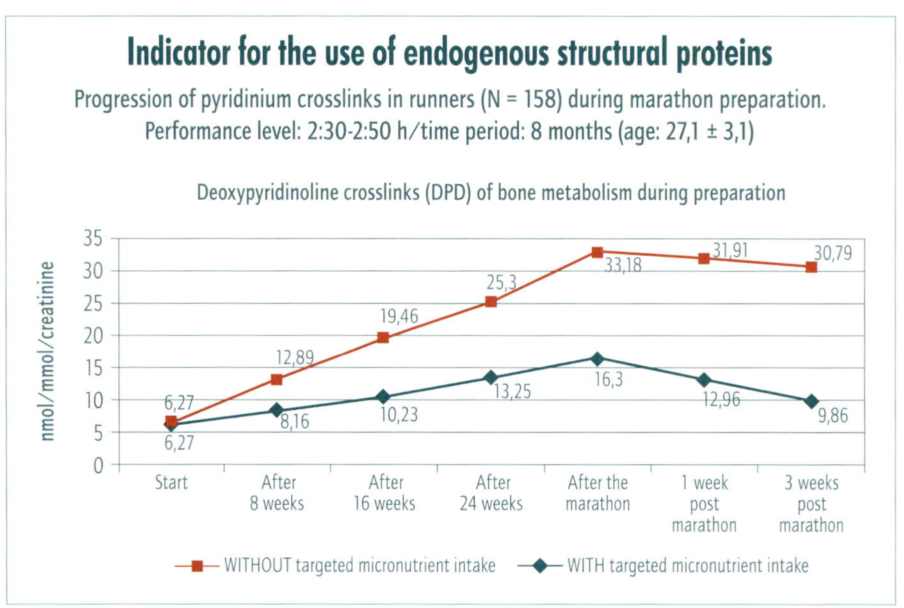

Fig. 138: Progression of pyridinium crosslinks (DPD) with and without customized micronutrient intake in 158 runners during marathon preparation (performance level: 2:30-2:50 h)

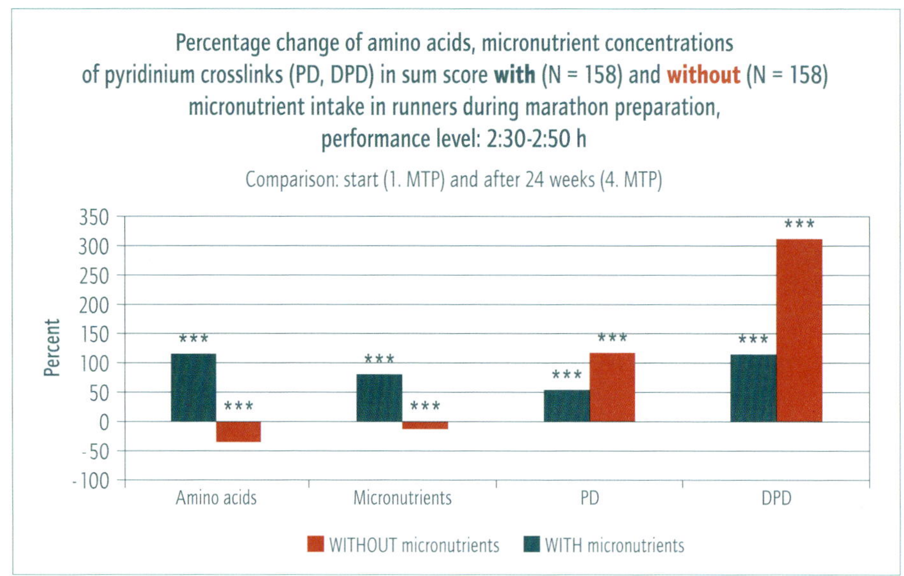

Fig. 139: Percentage change in amino acids, micronutrient concentrations and the pyridinium crosslinks (PD and DPD) with and without customized micronutrient intake in runners during marathon preparation (performance level: 2:30-2:50 h)

The pyridinium crosslinks show heavy use of endogenous structural proteins for the energy metabolism during the phase of unregulated intake of absent micronutrients. PD levels > 80 nmol/mmol creatinine and DPD levels > 15 nmol/mmol creatinine can verifiably have a negative effect on mental and physical performance capacity and increase the risk of injury.

Fig. 139 shows that the customized micronutrient intake in 158 runners led to a definite reduction in endogenous structural proteins.

6.3.8.3 Pro soccer players

The impact of optimized nutritional behavior and a targeted micronutrient formulation on the different parameters of the 48-hour HRV analysis was also ascertained in pro soccer players (N = 144) at various measuring time points over the course of the season with and without a targeted micronutrient formulation. Doing so made it possible to obtain detailed findings at different points during the season about the progression of different parameters of the vegetative nervous system.

The 48-hour HRV analysis always takes place from Tuesday to Thursday morning: before the start of preparation, at the end of preparation, after six weeks, at the end of the first leg and the end of the second leg. For the purpose of individualized analysis, the athletes were also given a special questionnaire which they completed. At the beginning of preparation, the test for homogeneity showed no statistically significant differences in the individual measured parameters. After six weeks of intensive preparation, the listed parameters show significant differences between the first MTP and the second MTP (tab. 33).

Tab. 33: Parameters of the 48-hour HRV measurement (M±SD) in 144 pro soccer players during a season without targeted micronutrient intake and during a season with targeted micronutrient intake. Percentage changes between the first and second MTP as well as key figures from the comparison test between the groups at the fourth MTP

Parameters		Before preparation start	preparation end after 6 weeks	End of first leg	End of second leg	Percentage change p < 0,001 Highly signifikant***	Comparison test between the groups
	Micro-nutrient intake	1. MTP	2. MTP	3. MTP	4. MTP	1. MTP → 4. MTP	1. MTP → 4. MTP
Stress index	without	362,9 ± 60,34	349,3 ± 62,52	352,53 ± 64,24	376,98 ± 64,92	3,95***	p < 0,001
	with	357,85 ± 55,88	178,85 ± 40,49	178,4 ± 47,08	171,35 ± 43,79	- 52,12***	p < 0,001
pNN50 (%)	without	7,23 ± 0,50	7,27 ± 1,15	7,35 ± 0,73	7,71 ± 0,69	6,63***	p < 0,001
	with	7,27 ± 0,61	12,32 ± 0,68	13,08 ± 0,79	13,81 ± 1,27	89,96***	p < 0,001
HF (ms^2)	without	88,90 ± 8,0	101,0 ± 9,05	102,76 ± 15,52	116,0 ± 16,0	31,46***	p < 0,001
	with	88,85 ± 6,69	501,5 ± 6,81	505,5 ± 16,65	513,65 ± 23,1	477,91***	p < 0,001
LF (ms^2) *	without	638,0 ± 82	651,0 ± 86	671,5 ± 88	704,0 ± 87,6	10,38***	p < 0,001
	with	644,0 ± 84,2	852,0 ± 72	847,24 ± 70,76	862,75 ± 70,28	33,99***	p < 0,001
LF/HF-ratio	without	7,28 ± 1,15	6,52 ± 1,05	6,70 ± 1,45	6,19 ± 1,21	- 14,94***	p < 0,001
	with	7,31 ± 0,61	1,70 ± 0,15	1,68 ± 0,18	1,78 ± 0,21	- 76,95***	p < 0,001

The stress index of 362,9 ± 60.34 at the first MTP prior to the start of the preparation phase in players without a targeted micronutrient formulation did not statistically change after six weeks at 349.3 ± 62.52, while the players with a targeted micronutrient formulation showed a highly significant ($p < 0,001$) reduction from 357.85 ± 55.88 to 178.85 ± 40.49 (fig. 140).

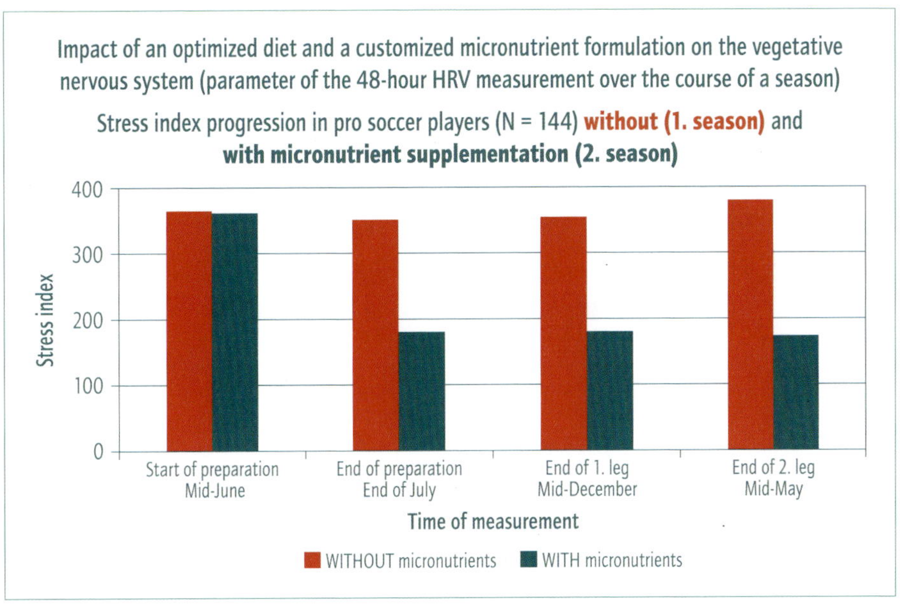

Fig. 140: Stress index before and after customized micronutrient intake in 144 pro soccer players

Players without a micronutrient formulation show a statistically higher stress index from the second to the fourth MTP than the players who then received a targeted micronutrient formulation during the second season. There are no discernable statistical differences in the stress index from the second to the fourth MTP. A low stress index verifiably indicates an improved stress tolerance. This is also apparent in the pNN50, which is considered a stable indicator for parasympathetic nervous system activity (fig. 141).

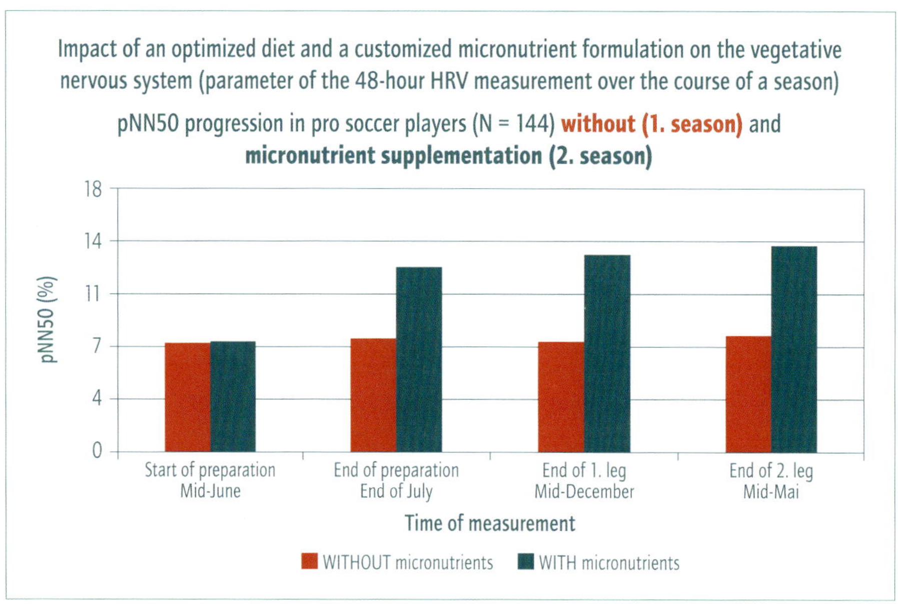

Fig. 141: pNN50 before and after customized micronutrient intake in 144 pro soccer players

The first MTP with a pNN50 of 7.23 ± 0.50 did not change at the second MTP with 7.27 ± 1.15 after six weeks in players without a micronutrient formulation and shows a disbalance of the vegetative nervous system towards of severe sympathetic nervous system activation.

89.3% of players also consistently describe increased inner restlessness combined with very poor sleep behavior, mental and physical overloading. The pNN50 from the second to the fourth MTP changes insignificantly.

During the second phase with a targeted micronutrient formulation the players show a highly significant increase ($p < 0.001$) from the first MTP with a pNN50 of 7.27 ± 0.61 to a pNN50 of 12.32 ± 0.68 after six weeks of an intensive preparation phase.

The results clearly reflect the players' subjective descriptions on the accompanying survey. They describe a balanced vegetative nervous system with good mental and physical performance capacity. The nightly sleep phases also show verifiably reduced sympathetic nervous system activity. The pNN50 also changes insignificantly at the additional MTPs at the end of the first and second leg and shows a well-balanced vegetative nervous system

This is also confirmed by the vegetative quotients (LF/HF-ratio; tab. 34), the balance between the sympathetic and parasympathetic nervous systems (fig. 142).

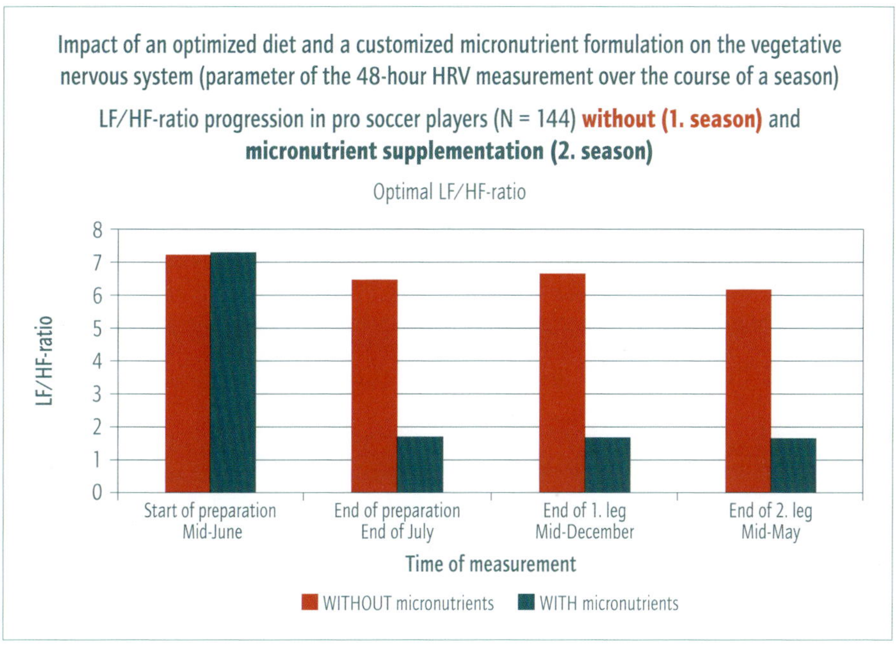

Fig. 142: Vegetative quotient before and after customized micronutrient intake in 144 pro soccer players

Conclusion and progression of pyridinium crosslinks (indicator for the use of endogenous structural proteins)

Both season assessments capture the players that optimized their nutritional behavior over the course of a 2-year evaluation period and also received their customized micronutrient formulation during the second season. The players show only 9% infection related, and 7% injury related absences as overload reactions of stressed connective tissue structures (muscle, tendons, ligaments). By comparison, during the first season, without a micronutrient formulation, the same players show 53% infection related absences and 45% injury related training interruptions of 12 weeks throughout.

The results of the pyridinium crosslinks (DPD) reveal that the pro players show no major use of endogenous structural proteins during the season with customized micronutrient intake (fig. 143.2). PD levels >80 nmol/mmol creatinine and DPD levels >15 nmol/mmol creatinine can verifiably have a negative effect on mental and physical performance capacity and increase the risk of injury. The progression of DPD levels is shown as an example. Further details about pyridinium crosslinks can be found on page 146-148.

The pro soccer players with micronutrient supplementation showed a significant improvement in intracellularly measured micronutrient concentrations (tab. 34 and fig. 143).

Tab. 34: Cellular micronutrient concentrations (M ± SD) of 144 pro soccer players during a season without targeted micronutrient supplementation and during a season with targeted micronutrient supplementation. Percentage change between the first and fourth MTP as well as key figures of the comparison test between the groups at the fourth MTP.

Parameter	Start of preparation Mid-Junei (1. MTP)		End of preparation End of July (2. MTP)		End of first leg Mid-December (3. MTP)		End of second leg Mid-May (4. MTP)		Percentage change 1. MTP 4. MTP Highly significant*** $p < 0,001$	
	without	with	without	with	without	with	without	with	without	with
Mg (mg/l ery.)	42,25 ± 3,74	44,10 ± 3,33	39,35 ± 3,04	48,50 ± 2,99	38.45 ± 3,25	54,50 +-3,53	37,94 ± 3,24	54,50 ± 3,53	-10,20***	23,60***
Zn (mg/l ery.)	11,33 ± 0,56	11,10 ± 0,77	9,57 ± 0,76	13,00 ± 0,40	9,25 ± 0,59	14,40 ± 0,64	8,87 ± 0,93	14,40 ± 0,64	-21,63***	28,71***
Se (µg/l ery.)	81,84 ± 7,86	82,01 ± 8,00	77,84 ± 8,01	119,75 ± 7,63	68,64 ± 7,13	138,75 ± 7,36	8,87 ± 0,93	138,75 ± 7,35	-26,88***	69,71***
Vit. B_1 (µg/l ery.)	61,26 ± 4,27	61,70 ± 5,11	56,36 ± 5,26	84,70 ± 4,98	53,0 ± 5,28	94,36 ± 4,61	48,36 ± 5,98	94,36 ± 4,61	-21,06***	53,03***
Vit. B_2 (µg/l ery.)	170,20 ± 20,12	172,22 ± 22,95	159,4 ± 18,81	379,10 ± 19,12	115,5 ± 24,07	458,50 ± 17,36	76,7 ± 20,15	458,50 ± 17,36	-55,48***	169,05***
Vit. B_9 (µg/l ery.)	557,0 ± 105	626,10 ± 98,9	524,10 ± 82,75	1.015,38 ± 97,65	501,46 ± 73,56	1.083,0 ± 100	396,02 ± 94,07	1083,0 ± 100	-28,96***	72,95***
HoloTC (pmol/l)	78,59 ± 11,42	67,60 ± 10,96	44,59 ± 13,4	132,0 ± 11,80	26,09 ± 10,71	156,20 ± 11,25	23,09 ± 13,41	156,20 ± 11,25	-70,62***	131,07***

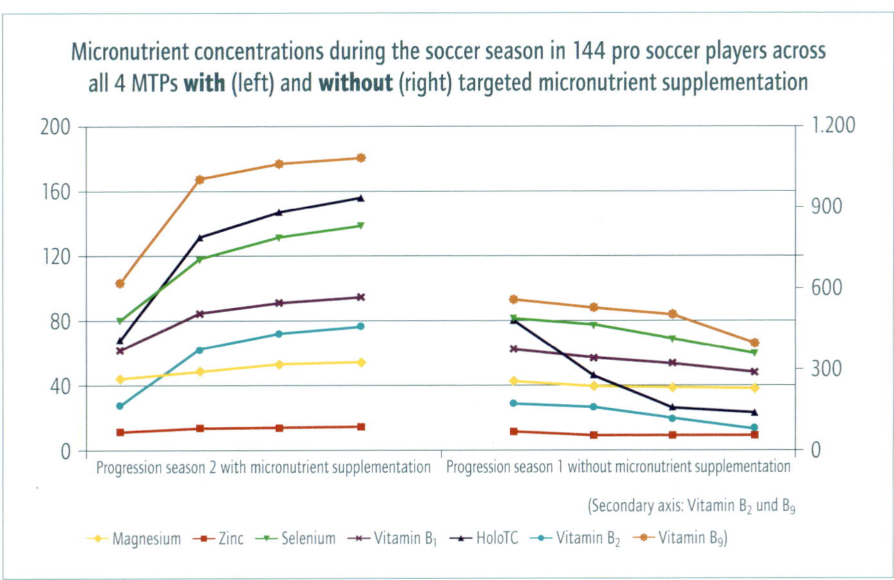

Fig. 143.1: Progression of cellular micronutrient concentrations (M ± SD) in 144 pro soccer players at different MTPs during one season (before the start of preparation, end of preparation after six weeks, end of the first leg, end of the second leg) without targeted micronutrient supplementation and during one season with targeted micronutrient supplementation

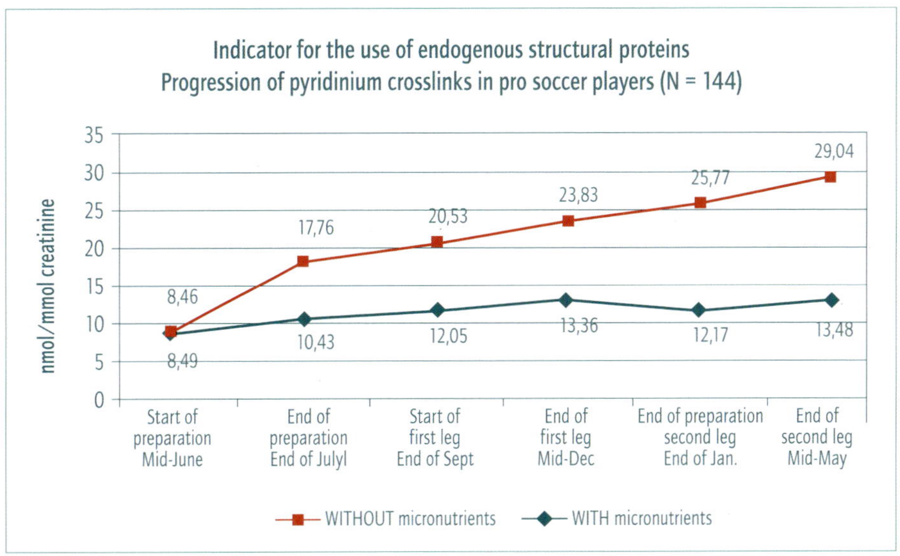

Fig. 143.2: Progression of pyridinium crosslinks (DPD) of the bone metabolism with and without customized micronutrient intake in 144 pro soccer players within a 2-year period

We could show comparable results with respect to the amino acid supply (tab. 35 and tab. 36 as well as fig. 144 and fig.145).

Tab 35: Progression of amino acid concentrations (M ± SD) in 144 pro soccer players at different MTPs during one season (before the start of preparation, end of preparation after six weeks, end of the first leg, end of the second leg) without targeted micronutrient supplementation and during one season with targeted micronutrient supplementation

Amino acid (mg/dl)	Start of preparation Mid-June (1. MTP)		End of preparation End of July (2. MTP)		End of first leg Mid-December (3. MTP)		End of second leg Mid-May (4. MTP)		Percentage change (1. MTP → 4. MTP) p < 0,001 Highly significant***		Comparison test between the groups 1. → 4. MTP
	without	with	without	with	without	with	without	with	without	with	
Arginine	0,86 -0,22	0,83 -0,11	0,66 -0,17	2,18 -0,34	0,50 -0,15	2,43 -0,12	0,40 -0,12	2,76 -0,62	-52,75***	231,73***	p < 0,001
Proline	3,09 -0,64	3,03 -0,89	2,74 -0,51	4,61 -0,96	2,53 0,55	4,91 -1,12	2,38 -0,45	5,14 1,24	-22,75***	69,64***	p < 0,001
Methionine	0,63 -0,13	0,57 -0,23	0,42 ±0,18	0,91 -0,19	0,39 -0,23	1,12 -0,32	0,15 -0,14	1,20 -0,41	-76,75***	110,00***	p < 0,001
Glycine	1,99 ± 0,13	1,96 -0,22	1,61 -0,31	2,91 -0,49	1,39 -0,24	3,24 -0,24	1,27 -0,28	3,51 -1,01	-36,34***	79,08***	p < 0,001
Glutamine	6,83 -0,87	7,01 -0,71	5,85 -0,88	9,18 -0,79	5,44 -0,87	10,59 -0,80	5,19 -0,68	11,05 -0,79	-23,93***	57,74***	p < 0,001
Glutamic acid	3,94 -0,47	3,84 -0,36	3,04 -0,61	6,16 -1,47	2,28 -0,54	7,50 -1,68	1,82 -0,44	9,14 -2,57	-53,71***	137,97***	p = 0,001
Leucine	2,31 -0,38	2,26 -0,51	1,97 -0,42	3,67 -0,67	1,77 -0,52	3,98 -0,89	1,63 -0,55	4,12 -1,36	-29,57***	82,30***	p = 0,001
Isoleucine	1,35 -0,28	1,16 -0,45	1,12 -0,33	2,49 -0,65	0,94 -0,17	2,73 -0,96	0,81 -0,21	2,91 -1,27	-40,25***	150,86***	p = 0,001
Valine	3,22 -0,39	3,30 -0,59	2,98 -0,51	3,82 -0,69	2,85 -0,55	4,09 -0,39	2,73 -0,32	4,21 -1,12	-15,16***	27,58***	p = 0,001

Tab. 36: Progression of amino acid concentrations (M ± SD) in 144 pro soccer players at different MTPs during one season (first MTP: before the start of preparation, end of preparation after six weeks, end of the first leg, end of the second leg) without targeted micronutrient supplementation and during one season with targeted micronutrient supplementation

Amino acid (mg/dl)	Start of preparation Mid-June (1. MTP)		End of preparation End of July (2. MTP)		End of first leg Mid-December (3. MTP)		End of second leg Mid-May (4. MTP)		Percentage change (1. MTP → 4. MTP) p < 0,001 Highly significant***		Comparison test between the groups 1. → 4. MTP
	without	with	without	with	without	with	without	with	without	with	
Alanine	2,73 -0,34	2,93 -0,50	2,52 -0,58	4,65 -0,96	2,32 -0,29	4,87 -1,33	2,16 -0,42	5,38 -1,51	-20,88***	83,62***	p < 0,001
Aspartic acid	0,26 -0,05	0,24 -0,11	0,16 -0,09	0,67 -0,13	0,13 -0,13	0,81 -0,25	0,11 -0,07	0,92 -0,37	-56,03***	282,92***	p < 0,001
Ornithine	1,00 -0,21	0,98 -0,44	0,87 -0,32	1,65 -0,56	0,76 -0,21	1,89 -0,85	0,62 -0,29	2,01 -0,95	-37,47***	105,10***	p < 0,001
Taurine	1,25 -0,39	1,29 -0,44	0,91 -0,29	2,62 -0,74	0,67 -0,31	2,89 -0,95	0,57 -0,24	3,06 -1,11	-54,65***	137,21***	p < 0,001
Histidine	1,61 -0,39	1,54 -0,49	1,24 -0,29	2,49 -0,51	1,03 -0,18	2,81 -0,87	0,91 -0,39	2,96 -1,23	-43,42***	92,21***	p < 0,001
Lysine	3,13 -0,52	3,12 -0,32	2,76 ± 0,42	3,95 -0,73	2,48 -0,50	4,35 -0,92	2,31 -0,46	4,51 -1,32	-26,12***	44,58***	p = 0,001
Phenylalanine	1,20 -0,27	1,12 -0,14	1,00 -0,17	2,54 -0,34	0,85 -0,31	2,75 -0,71	0,74 -0,24	2,94 -1,16	-38,72***	163,44***	p = 0,001
Tryptophan	1,32 -0,37	1,27 -0,22	1,18 -0,27	1,61 -0,41	0,97 -0,31	1,82 -0,69	0,84 -0,38	1,99 -0,89	-36,63***	56,69***	p = 0,001
Tyrosine	1,33 -0,36	1,40 -0,27	1,02 -0,28	2,38 -0,46	0,90 -0,30	3,24 -0,75	0,69 -0,23	3,45 -1,21	-48,50***	146,92***	p = 0,001

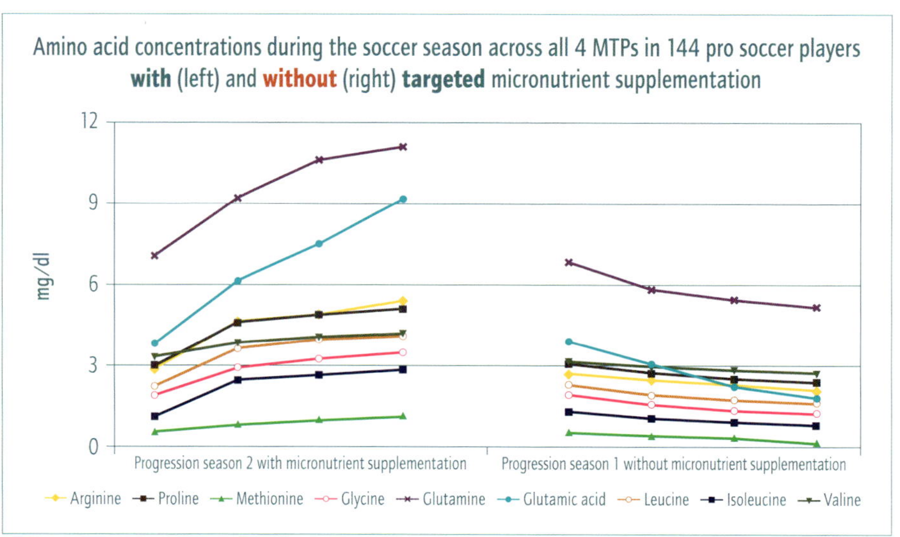

Fig. 144: Progression of amino acid concentrations (M ± SD) in 144 pro soccer players at different MTPs during one season (before the start of preparation, end of preparation after six weeks, end of the first leg, end of the second leg) without targeted micronutrient supplementation and during one season with targeted micronutrient supplementation

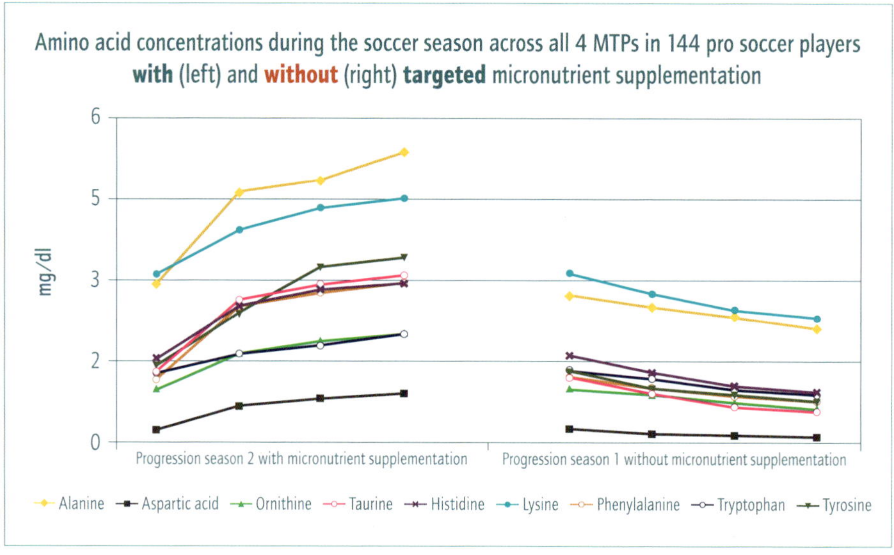

Fig. 145: Progression of amino acid concentrations (M ± SD) in 144 pro soccer players at different MTPs during one season (before the start of preparation, end of preparation after six weeks, end of the first leg, end of the second leg) without targeted micronutrient supplementation and during one season with targeted micronutrient supplementation

The progression of individual amino acid concentrations at the various measuring time points shows a highly significant decrease ($p < 0{,}001$) from the first to the second MTP in the pro players that did not receive a customized micronutrient formulation during the first season but demonstrate optimized nutritional behavior. Compared to the second season, the amino acid concentrations show a highly significant increase ($p > 0{,}001$) over the course of the season in the pro players with a targeted micronutrient formulation.

The relationship between the individual parameters of the 48-hour HRV measurement and the good cellular level of micronutrient concentrations as well as the progression of amino acid concentrations is scientifically proven by the results of the study. The pro soccer players with a targeted micronutrient formulation show a highly significant change in the individual parameters of the 48-hour HRV measurement, which has verifiably led to the improved balance of the vegetative nervous system (fig. 146).

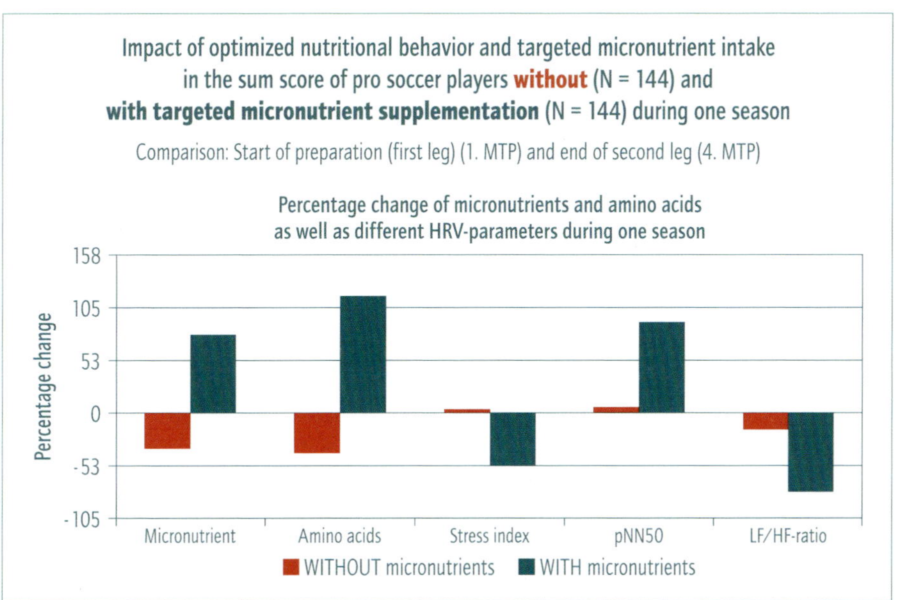

Fig. 146: Percentage change of amino acids, micronutrient concentrations with and without customized micronutrient intake in 144 pro soccer players

APPENDIX

1 Tables

An overview of the individual amino acids, micronutrients, and vitamins with respect to dosage and effect can be found in tables 37, 38, and 39.

Tab. 37: Recommended dosages of amino acids in the sense of customized micronutrient formulations, based on the results of our intracellular blood and urine analyses – functional energy metabolism, retrospective evidence-based studies, the globally unparalleled data base by SALUTO with more than 60,000 individuals tested, with those of the German Nutrition Society which are also German Association for Sports Medicine (DGSP) guidelines. An interpretation of the micronutrient analyses follows not because of the individual element concentrations but due to the deviation from the respective median value of the individual target reference values.

Amino acids	Important metabolic function	Recommended intake/ day (GNS: adults) *3	Food sources
Arginine **Semi-essential**	Energy metabolism Regeneration after stress and strain, important for the building of connective tissue Stimulation of the immune system, release of HGH	–	Fish Meat Soy products Whole-grain rice Peanuts Almonds
Glutamine **Non-essential**	Immune stability Energy production Stabilization of the blood sugar level Regeneration	–	Whole-grain products Soy products Dairy products
Glycine **Non-essential**	Stimulation of the immune system Building of immunoglobulins Liver detoxification processes	–	Beef Gelatin
Leucine/Isoleucine/ Valine **Essential branched-chain amino acids**	Energy production Protection from premature fatigue Regeneration	Leucine: 14 mg/kg bodyweight Isoleucine: 10 mg/kg bodyweight Valine: 14 mg/kg bodyweight	Legumes Oats Potatoes Whole-grain products Parmesan Eggs, meat, rice, dairy products

Possible symptoms of an undersupply	Optimal micronutrient concentrations via blood analyses Serum/plasma*2	Recommended dosages: Patients with different medical conditions *1	Recommended dosages: (Elite) competitive athletes *2
Weak immune system (increased infections) Increased injury risk to many connective tissue structures (tendons, ligament apparatus, etc.)	Serum/plasma: Deficient: < 2.2 mg/dl Optimal: > 20% above the median value	3-5g *1	3-8 g
Decreased physical performance capacity Increased tendency to infection Disturbed gut function	Serum/plasma: Deficient: < 9.8 mg/dl Optimal: > 20% above the median value	2-5 g *1	4-10 g
Impaired immune system Increased injury risk to many connective tissue structures (tendons, ligaments, etc.)	Serum/plasma: Deficient: < 2.8 mg/dl Optimal: > 20% above the median value	6-10 g *1	6-14 g
Quick physical and mental fatigue Muscle weakness Impaired ammonia detoxification	Serum/plasma: Deficient: Leucine: < 2.8 mg/dl Isoleucine: < 2.2 mg/dl Valine: < 4.1 mg/dl Optimal: > 20% above the median value	Leucine: 2 g Isoleucine: 1.5 g Valine: 2 g	Leucine 2-4 g Isoleucine: 1.5-3 g Valine: 2-4; 4 g

Amino acids	Important metabolic function	Recommended intake/ day (GNS: adults) *3	Food sources
Lysine Essential	Stimulation of the immune system Collagen synthesis Carnitine synthesis Building of connective tissue structures	10 mg/kg bodyweight	Potatoes Wheat Soy products Eggs (egg white) Meat
Methionine **Essential** **Sulfurous amino acid**	Building of many connective tissue structures (ligaments, tendons, etc.) Formation of glutathione (together with cysteine) Building of coenzyme Q_{10} for the synthesis of CP stores	13 mg/kg bodyweight	Lentils Soy products Skim-milk products Eggs Fish
Tryptophan Essential	Formation of hormones serotonin, melatonin, and the vitamin Niacin	3 mg/kg bodyweight	Bananas Soy Lentils Peanuts, parmesan Dairy products
Cysteine **Semi-essential,** **sulfurous amino acid**	Immunocompetence Cell protection from free radicals Glutathione synthesis (together with methionine)	–	Corn kernels Oats Eggs (egg white) Green vegetables (spinach, broccoli, etc.) Asparagus
Taurine (Amino-acid-like) Semi-essential	Stabilization of the immune system Strong antioxidant	–	Giblets Meat extracts Meat

*1 Customized micronutrient formulations are based on results from an intracellular blood test, additionally measured blood parameters, results from the nutritional analysis, and SALUTO's comprehensive, globally unparalleled database.

*2 Dosages that have verifiably significantly reduced the athletes' risk of infection and injury.

*3 The previously recommended intake is only derived from basic considerations on the relationship between energy consumption and an increased nutrient requirement. The listed recommendations are joint recommendations from German-speaking nutrition societies (Germany, Austria, Switzerland). Largely missing to date is sports-medicinal exercise-physiological data for the practical verification of the clinical relevance of previous dosage recommendations.

Possible symptoms of an undersupply	Optimal micronutrient concentrations via blood analyses Serum/plasma*2	Recommended dosages: Patients with different medical conditions *1	Recommended dosages: (Elite) competitive athletes *2
Increased susceptibility to infection Increased susceptibility to injury of many connective tissue structures (tendons, ligaments, etc.) Disruptions to the carnitine and fat metabolism	Serum/plasma: Deficient: < 3.4 mg/dl Optimal: > 20% above the target reference values	1 g *1	Up to 2.5 g
Worsening immune function Increased susceptibility to injury of many connective tissue structures Worsening regeneration after sprint loads	Serum/plasma: Deficient: < 0.9 mg/dl Optimal: > 20% above the target reference values	0,2 g *1	Up to 0.4 g
Sleep disturbances Bad mood Depressive mood	Serum/plasma: Deficient: < 2.0 mg/dl Optimal: > 20% above the target reference values	3 x 50-100 mg *1 5-HTP for pain relief, taken separately throughout the day	200-400 mg 5-HTP taken separately in the evening 2 hours before bed
Impaired immune function Oxidative stress Building of many connective structures (tendons, ligaments, etc.)	Serum/plasma: Deficient: < 1.8 mg/dl Optimal: > 0% above the target reference values	0.5-1 g cysteine as N-acetylcysteine *4 taken separately as needed	1-2 g cysteine as N-acetylcysteine *4 taken separately as needed
Impaired immune function Oxidative stress	Serum/plasma: Deficient: < 3.2 mg/dl Optimal: > 20% above the target reference values	0.2-0.4 g *1 To date we have not administered additional taurine	0.4-0.8 g To date we have not administered additional taurine

*4 In athletes with a particular skin problem (such as acne, neurodermatitis, psoriasis) an increase of certain B vitamins can cause the complexion to worsen. General remark: Targeted intravenous treatments in cases of exhaustion are not taken into account here, but in some cases can be performed by consulting physicians after specific indication.

*5 Vitamins A and D should not exceed the listed reference values.

*6 Specifications by Gröber from "Metabolic Tuning instead of Doping" Hirzel Publishing, scientific publishing company Stuttgart/Germany (3. edition).

Tab 38: Recommended dosages for micronutrients in the sense of individual micronutrient formulations, based on the results of our intracellular blood and urine analyses – functional energy metabolism, retrospective evidence-based studies, the globally unparalleled data base by SALUTO with more than 60,000 individuals tested, with those of the German Nutrition Society which are also German Association for Sports Medicine (DGSP) guidelines. An interpretation of the micronutrient analyses follows not because of the individual element concentrations, but due to the deviation from the respective median value of the individual target reference values.

Vitaminoids	Important metabolic function *3	Recommended dosage/day (GNS: adults) *3	Food sources
L-carnitine	Energy generation from fats (mitochondria and fatty acid transport)	Approx. 100 mg (GNS: no recommendation)	Lamb Beef
Coenzyme Q_{10}	Antioxidant Stabilization of cell membrane Cellular energy production	Unknown	Olive oil Eggs Liver
Alpha-lipoic acid	Energy generation from carbohydrates (in conjunction with B_1)	Unknown	–
Omega-3 fatty acids	Building block for every cell membrane Immune system stabilizing Anti-inflammatory Increases mental performance and concentration capacity Improved oxygen supply to organs	Unknown	Oil from saltwater fish (herring, wild salmon) Soybean oil Corn oil Fish oil generally contains about 30-35% Omega-3 fatty acid

*1 Customized micronutrient formulations are based on results from an intracellular blood test, additionally measured blood parameters, results from the nutritional analysis, and SALUTO's comprehensive, globally unparalleled database.

*2 Dosages that have verifiably significantly reduced the athletes' risk of infection and injury.

*3 The previously recommended intake is only derived from basic considerations on the relationship between energy consumption and an increased nutrient requirement. The listed recommendations are joint recommendations from German-speaking nutrition societies (Germany, Austria, Switzerland). Largely missing to date is sports-medicinal exercise-physiological data for the practical verification of the clinical relevance of previous dosage recommendations.

Possible symptoms of an undersupply	Optimal micronutrient concentrations via blood analyses Serum/plasma *1, 4 Whole blood, intracellular	Recommended dosages: Patients with different medical conditions *1	Recommended dosages: Competitive and elite athletes *2
Impaired fat and carbohydrate metabolism Susceptibility to infection Fatigue	Free carnitine (FC) in plasma: 40-60 µmol/l Ratio of acylcarnitine/ free carnitine AC/FC quotient: Normal or good supply: Fasting: < 0,7 Postprandial: < 0,4	750-2.000 g *1	1.000-3.000 mg
Still unknown	In whole blood: Deficient: > 1,5 mg/l Cholesterol corrected: < 0,2 µmol/mmol Optimal: 20% above target-/reference values	60-300 mg *1	30-150 mg
Still unknown	–	60-200 mg *1	200-400 mg
Low stress tolerance Low mental performance and concentration capacity Increased risk of damage to cell membranes Increased susceptibility to infection	Optimal diagnostics Membrane lipids in ery. Serum: Alpha-linolenic acid: Deficient: < 30 mg/l Deficient EPA: < 30 mg/l DHA: < 110 mg/l Optimal: 20% above target reference values	1,0-5 g *1 Reference: 1 g (Omega-3 fa = approx. 3 g fish oil) (High concentrates contain 85%)	3-7 g Reference: 1 g (Omega-3 fa = approx. 3 g fish oil) (High concentrates contain 85%)

*4 In athletes with a particular skin problem (such as acne, neurodermatitis, psoriasis) an increase of certain B vitamins can cause the complexion to worsen. General remark: Targeted intravenous treatments in cases of exhaustion are not taken into account here, but in some cases can be performed by consulting physicians after specific indication.

*5 Vitamins A and D should not exceed the listed reference values.

*6 Specifications by Gröber from "Metabolic Tuning instead of Doping"- Hirzel Publishing, scientific publishing company Stuttgart/Germany (3. edition).

THE POWER OF MICRONUTRIENTS

Minerals	Important metabolic function	Recommended dosage/day (GNS: adults) *3	Food sources
Calcium	Stabilization of cell membranes Plays an important role in cardiac-neural-muscle function Stabilization of bony structure	1.000-1.200 mg	Dairy products Kale
Magnesium	Carbohydrate, fat, and protein metabolism Activation of more than 300 enzymes (important for energy metabolism) Stabilization of cell membranes Calcium antagonist	300-400 mg	Certain mineral waters Whole grains Bananas Nuts
Potassium	Short-term replenishment of carbohydrate stores in the liver and musculature Excitation of nerves and muscles Conduction of stimuli in the heart muscle	2.000 mg (estimated value for minimum intake)	Bananas Potatoes Vegetables 100% fresh-pressed juices Wheat germ
Sodium Chloride	Regulation of water and electrolyte balance Preservation of membrane potential	2.000 mg sodium 3.000 mg chloride	Table salt Mineral water
Iron	Oxygen supply to organs, hematopoiesis Enzyme component (for energy supply)	10-15 mg	Meat Chanterelle mushrooms Unsulfurated apricots Blueberry extract

Possible symptoms of an undersupply	Optimal micronutrient concentrations via blood analyses Serum/plasma *4, intracellular *1	Recommended dosages: Patients with different medical conditions *1	Recommended dosages: Competitive and elite athletes *2
Increased nervousness Muscle twitching Cramping (tetany)	Calcium status: *4 Serum concentration 2.2-2.65 mmol/l (= 8.8-10.6 mg/dl)	200-400 mg	400-700 mg
Increased susceptibility to feeling stressed Hyperexcitability Worsening regeneration Cramping Eyelid and muscle twitching No optimal development of endurance capacity	Intracellular (in ery.): *1 Deficient: < 50 mg/l ery. Optimal: 20% above the median value	200-700 mg *1 spread out over the course of the day e.g. magnesium bisglycinate does not lead to gastrointestinal problems	300-800 mg: Optimal for athletes: Magnesium biclycinate spread out over the course of the day Does not cause gastrointestinal problems
Muscle weakness Lack of appetite Worsening regeneration Cardiac arrythmias	Optimal intracellular: *1 Technically hardly feasible as the blood can only be transported in a frozen state Alternative: whole blood Deficient: < 1.798 mg/l Optimal: 20% above the target-reference values	200-500 mg	500-1000 mg Particularly for endurance athletes during the regeneration phase
Increased proneness to cramping Insufficient regeneration after athletic exertion Dizziness	Serum/plasma: *4 136-145 mmol/l	Low-sodium diet with high blood pressure	3000-6000 mg sodium 4500-9000 mg chloride
Fatigue Poor regeneration prevents optimal development of endurance capacity Dizziness Pale skin Nervousness Alopecia	Serum-ferritin: *1 Optimal for competitive athletes Men: > 120 µg/l Woman: > 60 µg/l	10-100 mg *1 Should only be taken after blood test and on doctor's orders Iron infusion if necessary	10-100 mg *1 Should only be taken after blood test and on doctor's orders Iron infusion if necessary

Trace elements	Important metabolic function in sports	Recommended intake/day (GNS: adults) *3	Food sources
Iodine	Energy balance Production of thyroid hormones	150-200 μg	Saltwater fish Iodized salt
Zinc	Immune stabilization Supports many functions in the carbohydrate, fat, and protein metabolism Insulin storage	7-10 mg	Oysters Cheese Legumes Filet of beef
Selenium	Regulation of thyroid hormones Antioxidant Immune stability	30-70 μg (estimated value for an appropriate intake)	Whole-grain (products) Saltwater fish Liver
Chrome	Fat metabolism Optimizes replenishment of glycogen stores	30-100 μg	Meat Cheese Mushrooms Brewer's yeast
Copper	Iron transport Immune stability Hematopoiesis Collagen and neurotransmitter synthesis	1-1,5 mg	Nuts
Manganese	Carbohydrate metabolism Antioxidant Bone and cartilage growth	2-5 mg	–

Possible symptoms of an undersupply	Optimal micronutrient concentrations via blood analyses Serum/plasma *4, intracellular *1	Recommended dosages: Patients with different medical conditions *1	Recommended dosages: Competitive and elite athletes *2
Performance problems Night sweats Increasing fatigue Lack of motivation Trouble concentrating Poor regeneration after stress and strains Goiter formation	JIodine status: excretion in urine between 100-150 µg/d > 100 µg/g creatinine	50-200 µg (no iodine should be taken when TSH-level is < 1,3 µIU/l, because the stress index is significantly elevated)	100-400 µg (begin with a targeted iodine intake if there is no load-induced TSH elevation > 2.5 µIU/l)
Allergic reactions: Worsening pollen allergy Increasing susceptibility to infection Alopecia Impaired sense of smell and taste	Intracellular (in ery.) *1 Deficient: < 14 mg/l Ery. Athletes with a pollen allergy < 15 mg/l ery. Optimal: > 20 % above the target-reference values	10-45 mg *1 short-term, more during active infections	20-50 mg short-term, more during active infections
Increased susceptibility to infection Impaired thyroid function	Intracellular (in ery.) *1 Deficient: < 114,4 µg/l ery. Optimal: > 20 % 20% above the target-reference values	50-250 µg	100-200 µg Taking more than 200 µg long-term can have a toxic effect
Impaired glucose utilization in energy generation and fat metabolism	Whole-blood status: *2, 4 < 90 µmol/l Serum: 13 mmol/l Standard values show significant differences Optimal: 20% above the target-reference	100-500 µg *1	200-600 µg
Higher infection rate Anemia Fat metabolism Brittle bones	Copper status: serum: *4 Men: 80-130 mg/dl Woman: 75-120 µg/dl (Urine: 10 bis µg /24 h) The serum copper level is often elevated during acute and chronic infections Whole-blood status: *2 Deficient: < 0,7 mg/l Optimal: > 20% above the median value	2-4 mg	4-8 mg Optimal ratio of zinc/copper is an important prerequisite for the proper resorption from the gastrointestinal tract
Impaired carbohydrate metabolism Inhibited bone and cartilage growth	Reference range: Deficient: Whole blood:: < 7,0 µg/l Optimal: 20 % above the median value	5-10 mg	10-20 mg

Tab 39: Recommended dosages for vitamins in the sense of individual micronutrient formulations, based on the results of our intracellular blood and urine analyses – functional energy metabolism, retrospective evidence-based studies, the globally unparalleled data base by SALUTO with more than 60,000 individuals tested, with those of the German Nutrition Society which are also German Association for Sports Medicine (DGSP) guidelines. An interpretation of the micronutrient analyses follows not because of the individual element concentrations, but due to the deviation from the respective median value of the individual target reference values.

Vitamins	Important metabolic function	Recommended intake/day (GNS: adults) *3	Food sources
Vitamin C Ascorbic acid Water soluble	Carnitine biosynthesis Stress resistance Neurotransmitter balance Antioxidative cell protection Iron utilization Protection of vessels (endothelium) Hormone metabolism	100 mg	Kiwi Broccoli Peppers
Vitamin B_1 Thiamine Water soluble	Mental performance capacity Energy-related utilization of carbohydrates (mitochondria) Heart, nerve, and muscle metabolism	1.2 mg In athletes: 0.5 mg/1000 kcal energy turnover	Cereal germ Soy flour Pork Bran Vegetables Nuts Giblets
Vitamin B_2 Riboflavin Water soluble	Antioxidative cell protection Stabilization of immune system	1.4 mg In athletes: 0.6 mg/1000 kcal energy turnover	Wheat germ Wheat bran Dairy products Grains
Vitamin B_3 Niacin/niacinamide Water soluble	Antioxidative cell protection Stabilization of immune system Mitochondrial energy metabolism	13-17 mg Niacin equivalent = 60 g tryptophan In athletes: 6.6 mg/1000 kcal energy turnover	Meat Fish Giblets Dairy products Grains

Possible symptoms of an undersupply	Optimal micronutrient concentration via blood tests Serum/plasma *6 Excretion/urine *6 Intracellular *1 (in erythrocytes)	Recommended dosages: Patients with different medical conditions *1	Recommended dosages: Competitive and elite athletes *1
Low stress resistance and regeneration Susceptibility to infection Muscle weakness Pain in limbs and joints Problems with motivation, concentration, and performance Irritability	Plasma/serum: Good supply: *1 > 90-170 µmol/l Elite athletes: *6 150-200 µmol/l	1.000-5.000 mg *1	500-3.000 mg (also intravenously in cases of infections and sports injuries)
Low stress resistance and regeneration Susceptibility to infection Muscle weakness Pain in limbs and joints Insomnia Irritability Loss of appetite Problems with motivation, concentration, and performance	Intracellular (in ery.) *1 Deficient: < 73 µg/l ery. Optimal: 20% above the respective target value	5-70 mg *1	10-70 mg *4
Susceptibility to infection Problems with motivation, concentration, and performance Muscle weakness	Intracellular (in ery.) *1 Deficient: < 380,1 µg Optimal: 20% above the respective target value	10-70 mg *1	20-70 mg *4
Susceptibility to infection Loss of appetite Muscle weakness Problems with motivation, concentration, and performance	Vitamin-B_3 status: *6 Ratio of renal excretion of N-methyl-2-pyridone-3-carboxamide and N-methyl nicotinamide in urine normal: 1.3-4.0; 1,3-4,0; Deficiency: < 1,0	10-70 mg *1	20-60 mg *4 Athletes can react with skin allergies to doses > 30 mg

Vitamins	Important metabolic function	Recommended intake/day (GNS: adults) *3	Food sources
Vitamin B_5 Pantothenic acid Water soluble	Carbohydrate, fat, and protein utilization, synthesis of stress hormones Availability of acetyl-coa (universal metabolic building block)	6-10 mg	Fish Mushrooms Legumes
Vitamin B_6 (Pyridoxin)	Immunocompetence Regulation of amino acid metabolism (Protein structure) Neurotransmitter synthesis Glycogen metabolism Homocysteine detoxification	1.4-1.5 mg in athletes: Depending on protein absorption: 0.15 mg/mg protein/day	Whole-grain products Yeast Fish Poultry
Vitamin B_9 Folic acid Water soluble	Energy metabolism Protein structure Cell regeneration (mucosa) DNA-building, nerve building and protection	0,4 mg Folate equivalent	Leafy green vegetables Whole grain
Vitamin B_{12} Cobalamin	Energy metabolism Nerve building and protection Hematopoiesis Cell regeneration DNA-building	3,0 µg	Liver Dairy products Fish Eggs
Biotin Vitamin H Water soluble	–	30-60 µg (estimated value for an appropriate intake)	Giblets Peas Nuts

Possible symptoms of an undersupply	Optimal micronutrient concentration via blood tests Serum/plasma *6 Excretion/urine *6 Intracellular *1 (in erythrocytes)	Recommended dosages: Patients with different medical conditions *1	Recommended dosages: Competitive and elite athletes *1
Low stress resistance and regeneration	Pantothenic acid status in urine *1, 6 Standard value: > 1 mg/d	10-70 mg	20-80 mg *4
Susceptibility to infection Low stress resistance and regeneration Muscle weakness Insomnia Loss of appetite Irritability Lack of motivation and concentration Performance problems	Intracellular (in ery.) Deficient: < 215 µg/l ery. Optimal: 20% above target value	5-60 mg	10-80 mg *4
Low stress resistance and regeneration Susceptibility to infection Muscle weakness Insomnia Irritability Lack of motivation and concentration Performance problems	Intracellular (in ery.) Deficient: < 796 µg/l ery. *1 Optimal: > 20% above mean value	0,4-2,0 g	0,8-1,6 g *4
Susceptibility to infection Lack of motivation and concentration Poor performance Irritability Muscle weakness Low stress resistance and regeneration Insomnia	Serum HoloTC Concentration: *1 Intracellular not measurable as fluctuation is too great Deficient: < 80 pmol/l Optimal: 20% above target value	20-60 µg	30-400 µg *4
Muscle aches Low stress tolerance and regeneration Lack of motivation and concentration Performance problems	Biotin excretion in urine: *6 25-50µg/24h	45-100 µg	45-200 µg

THE POWER OF MICRONUTRIENTS

Vitamins	Important metabolic function	Recommended intake/day (GNS: adults) *3	Food sources
Vitamin D *5 Calciferol Fat-soluble	Bone metabolism Calcium resorption and utilization Muscle contraction Strengthening the immune system	5-25 µg (200-2000 IU)	Saltwater fish Veal Mushrooms, egg yolk, butter, hard cheese To convert Vitamin D precursors: Sunlight
Vitamin A *5 Retinol Fat-soluble	Building and preservation of mucous membranes Strengthening immune defense Antioxidant Protective function Hormone metabolism	0,8-1 mg (2500-3500 IU)	Cod liver oil Giblets Eggs Dairy products
Vitamin E (alpha, gamma tocopherol etc.)	Antioxidant Protective function Regulation of the immune system and metabolic inflammation Vascular protection	12-15 mg Tocopherol equivalent or 18-22 IU (Estimated value for an appropriate intake)	Olive oil Wheat germ oil Walnuts
Vitamin K Phylloquinone	Bone metabolism Bone formation Blood formation and clotting	60-80 µg (Estimated value for an appropriate intake)	Leafy green vegetables Sauerkraut

*1 Customized micronutrient formulations are based on results from an intracellular blood test, additionally measured blood parameters, results from the nutritional analysis, and SALUTO's comprehensive, globally unparalleled database.

*2 Dosages that have verifiably significantly reduced the athletes' risk of infection and injury.

*3 The previously recommended intake is only derived from basic considerations on the relationship between energy consumption and an increased nutrient requirement. The listed recommendations are joint recommendations from German-speaking nutrition societies (Germany, Austria, Switzerland). Largely missing to date is sports-medicinal exercise-physiological data for the practical verification of the clinical relevance of previous dosage recommendations.

Possible symptoms of an undersupply	Optimal micronutrient concentration via blood tests Serum/plasma *6 Excretion/urine *6 Intracellular *1 (in erythrocytes)	Recommended dosages: Patients with different medical conditions *1	Recommended dosages: Competitive and elite athletes *1
Susceptibility to fatigue fractures Muscle weakness Susceptibility to infection Delayed regeneration Low bone density and load-bearing capacity	Vitamin-D-status: *1, 6 25-OH-Vitamin D_3 in serum: Deficient: < 130 nmol/l Optimal: 20% above target value	3.000-5.000 i.E. per day, particularly during fall and winter, additionally a good calcium supply is important	4000-8000 IU per day *1, particularly in fall and winter Athletes playing indoor sports should also ensure a good calcium supply
Susceptibility to infection Low resilience Delayed regeneration	Serum level (good supply): (gute Versorgung): *1 ,6 > 30 µg/dl Interpretation of serum levels is difficult due to the retinol homeostasis	0.8-1 mg 2500-3500 IU	1-3 mg *6 2500-10,000 IU Endurance athletes should not take more than 10,000 IU long-term. Beneficial to endurance athletes with frequent upper respiratory infections
Performance problems Muscle aches Risk of oxidative muscle damage Impaired regeneration ability Susceptibility to infection	Vitamin-E (alpha tocopherol) status: *1, 2, 6 Plasma concentration Deficient: < 38 µmol/l Optimal: 20% above target value	150-250 mg 275 IU-375 IU	200-400 mg 300 IU – 600 IU
–	Vitamin-K status *6 Plasma: standard value around 1,0 nmol/l	60-150 µg	150-500 µg (Use caution when taking blood thinners)

*4 In athletes with a particular skin problem (such as acne, neurodermatitis, psoriasis) an increase of certain B-vitamins can cause the complexion to worsen. General remark: Targeted intravenous treatments in cases of exhaustion are not taken into account here, but in some cases can be performed by consulting physicians after specific indication.

*5 Vitamins A and D should not exceed the listed reference values.

*6 Specifications by Gröber from "Metabolic Tuning instead of Doping"- Hirzel Publishing, scientific publishing company Stuttgart, Germany (3. edition).

2 Bibliography

- Bechthold, A., Albrecht, V., Leschik-Bonnet, E. & Heseker, H. (2012a). Beurteilung der Vitaminversorgung in Deutschland. Teil 1: Daten zur Vitaminzufuhr. *Ernährungs Umschau, 59,* 324-336.
- Bechthold, A., Albrecht, V., Leschik-Bonnet, E. & Heseker, H. (2012b). Beurteilung der Vitaminversorgung in Deutschland. Teil 2: Kritische Vitamine und Vitaminzufuhr in besonderen Lebenssituationen. *Ernährungs Umschau, 59*, 396-401.
- Biesalski, K., Bischoff, S. C. & Puchstein, C. (2010). *Ernährungsmedizin: Nach dem Curriculum Ernährungsmedizin der Bundesärztekammer.* (4. vollständig überarbeitete und erweiterte Auflage). Thieme, Stuttgart.
- Bischoff-Ferari, H. A., Giovannucci, E. & Willet, W. C. et al (2006). Estimation of optimal serum concentrations of 25-hydroxyvitamin D for multiple health outcomes. *Am J Clin Nutr, 84*, 18-28, https://doi.org/10.1093/ajcn/84.1.18.
- Blicharz, T. M., Gong, P. & Bunner, B. M. et al. (2018). Microneedle-based device for the one-step painless collection of capillary blood samples. *Nature biomedical engineering, 2*, 151-157.
- Bolland, M. J., Grey, A. & Avenell, A. (2018). Effects of vitamin D supplementation on musculoskeletal health: A systematic review, meta-analysis, and trial sequential analysis. *Lancet Diabetes & Endocrinology, 6,* 847-858. DOI:https://doi.org/10.1016/S2213-8587(18)30265-1.
- Eckert, N. (2018). Kardiovaskuläre Prävention: Rolle der Omega-3-Fettsäuren ist unklar. *Deutsches Ärzteblatt, 115 (9)*, A-388/B-327/C-327.
- Global Burden of Disease Study 2013 Collaborators (2015). Global, regional, and national incidence, prevalence, and years lived with disability for 301 acute and chronic diseases and injuries in 188 countries, 1990-2013: A systematic analysis for the Global Burden of Disease Study 2013. *Lancet, 386*, 743-800.
- Gröber, U. (2015). *Interaktionen: Arzneimittel und Mikronährstoffe.* (2. Auflage). Wissenschaftliche Verlagsgesellschaft, Stuttgart.
- Gröber, U. (2018). *Arzneimittel und Mikronährstoffe Medikationsorientierte Supplementierung.* (4. aktualisierte und erweiterte Auflage). Wissenschaftliche Verlagsgesellschaft, Stuttgart.

- Gronegger, I. (2015). *Schilddrüsen-Unterfunktion, Hashimoto und Hormone.* ISBN 978-1517358907; https://schilddruesen-unterfunktion.de
- Henz, D., Schöllhorn, W. & Poeggeler, B. (2018). Mobile phone chips reduce increases in EEG brain activity induced by mobile phone-emitted electromagnetic fields. Orginal research. *Frontiers of Neuroscience*, April 2018.
- Henz, D. (2019). Application of a bluetooth headset, cable headset, and a smartphone chip on the smartphone. Do these devices reduce effects on EEG brain activity induced by smartphone-emitted electromagnetic fields? *Psychophysiology, 56*, S 1, S 53.
- Herbst, R. (2020). Der Einfluss von Aminosäuren auf die Schmerzsymptomatik und das allgemeine Wohlempfinden. In A. Dreier, R. Merk & B. Seel (Hrsg.) *Meilensteine in der Gesundheitsmedizin.* (Schriftenreihe der FHM). Heft 12, 46-59.
- Holick, M. F. (2007). Vitamin D deficiency. *N Engl J Med, 357*, 266-281.
- Konofal, E., Lecendreux, M., Arnulf, I. & Mouren, M. C. (2004). Iron deficiency in children with attention-deficit/hyperactivity disorder. *Arch Pediatr Adolesc Med, 158*, 1113-1115; doi: 10.1001/archpedi.158.12.1113
- Kotsis, E., Kotsis, E. & Perrea, D. N. (2019). Vitamin D levels in children and adolescents with attention-deficit hyperactivity disorder (ADHD): A meta-analysis. *Atten Defic Hyperact Disord, 11*, 221-232; doi: 10.1007/s12402-018-0276-7
- Kuklinski, B. (2015). *Mitochondrien: Symptome, Diagnose und Therapie.* Aurum, Freiburg.
- Kuklinski, B. (2010). *Gesünder mit Mikronährstoffen.* Aurum, Freiburg.
- Liebram, C. & Nauber, T. (2015). Nur jeder zwanzigste Mensch ist wirklich gesund. *Die Welt* vom 09.06.2015; https://www.welt.de/gesundheit/article142167267/Nur-jeder-zwanzigste-Mensch-ist-wirklich-gesund.html
- Mosetter, K. et al. (2018). *Zucker der heimliche Killer. Mit dem 4-Schritte-Entwöhnungsprogramm raus aus der Zuckersucht.* 4. Aufl. GU, München.
- Mosetter, K. (2020). Organische Säuren im Stoffwechsel. In A. Dreier, R. Merk & B. Seel (Hrsg.), *Meilensteine in der Gesundheitsmedizin.* (Schriftenreihe der FHM). Heft 12, Seite 5-20.

- Müller, J. M. (2020). *Der Zusammenhang zwischen HS-Omega-3-Index und der Aufnahme von Mikronährstoffen.* – Eine retrospektive Interventionsstudie im Rahmen der Masterarbeit des MMA Mikronährstofftherapie/Regulationsmedizin an der FHM Bielefeld.
- Murr, C., Pilz, S. & Grammer, T. B. (2012). Vitamin D deficiency parallels inflammation and immune activation, the Ludwigshafen Risk and Cardiovascular Health (LURIC) study. *Clin Chem Lab Med, 50*, 2205-12. doi: 10.1515/cclm-2012-0157.
- Mursu, J., Robien, K. & Harnack, L. J. et al. (2011). Dietary supplements and mortality rate in older women. The Iowa Women's Health Study. *Arch Intern Med, 171*, 1625-1633. https://jamanetwork.com/journals/jamainternalmedicine/fullarticle/1105975
- Poeggeler, B. (2017). Ein Meilenstein in der Gesundheitsmedizin. *Meine Gesundheit, 33*, 12-13.
- Pollak, S. (2020). *Minimalinvasive Blutanalyse – eine Alternative für den Therapeuten.* Studien im Rahmen der SIP- und Zertifikatsarbeit im Rahmen der Masterarbeit des MMA Mikroährstofftherapie/Regulationsmedizin an der FHM Bielefeld.
- Raman, M. et al. (2011). Vitamin D and gastrointestinal diseases: inammatory bowel disease and colorectal cancer. *Ther Adv Gastroenterol, 4*, 49-63.
- Ross, J. (2017). *Was die Seele essen will.* (3. Aufl. Klett-Cotta). Stuttgart.
- Siegmann, E. M., Müller, H. H. O. & Luecke, C. et al. (2018). Asssociation of depression and anxiety disorders wit autoimmune thyroiditis. A systematic review and metaanalysis. *JAMA Psychiatry, 75*, 577-584.
- Smith, M. R. & Myers, S. S. (2018). Impact ofanthropogenic CO_2 emissions on global human nutrition. *Nature Climate Change, 8*, 834-839; Download Citation: ORCID:orid.org/0000-0001-5207-2370
- Stiftung für Mikronährstoffe, Prävention, Gesundheit, Lebensqualität (2015). *Abschlussbericht der Magnesium-Studie: Der Einfluss von oral supplementiertem Magnesium auf die intrazelluläre Magnesiumkonzentration, auf kardiovaskuläre Parameter auf die Lebensqualität, auf die Balance des vegetativen Nervensystems sowie auf die körperliche Leistungsfähigkeit* (unveröffentlicht).
- Stecco, C. (2018). *Functional atlas of the human fascial system.* Elsevier, Amsterdam.

- Strang, S., Hoeber, C. & Uhl, O. et al. (2017). Impact of nutrition on social decision making. *PNAS, 114, (25)*, 6510-6514; https://doi.org/10.1073/pnas.1620245114.
- Troesch, B., Eggerdorfer, M. & Weber, P. (2018). The role of vitamins in aging societies. *Int J Vitam Nutr, 82 (5)*, 355-9; OM & Ernährung 2018, Nr. 162.
- Tseng, P. T., Cheng, Y. S. & Yen, C. F. et al. (2018). Peripheral iron levels in children with attention-deficit hyperactivity disorder: a systematic review and meta-analysis. *Sci Rep, 8*, 788; doi: 10.1038/s41598-017-19096-x
- Wienecke, E., Ahlers, J. & Karamann, M. (2020). Stellenwert der Mikronährstoffe bei Kindern mit einer Aufmerksamkeits-/Hyperaktivitätsstörung. Mikronährstoffdiagnostik und -therapie bei 162 Patienten im Rahmen einer retrospektiven Untersuchung. In A. Dreier, R. Merk & B. Seel (Hrsg.). *Meilensteine in der Gesundheitsmedizin. (*Schriftenreihe der FHM). Heft 12, S. 60-71.
- Wienecke, E. (2020). Die Schilddrüse als Regulator der „inneren Balance" – Einfluss der Schilddrüse auf die psychophysische Leistungsfähigkeit anhand evidenzbasierter retrospektiver Studien. In A. Dreier, R. Merk & B. Seel (Hrsg.). *Meilensteine in der Gesundheitsmedizin. (*Schriftenreihe der FHM). Heft 12, S. 36-44.
- Wienecke, E. (2018). Die Schilddrüse als Regulator für eine optimale mentale und physische Leistungsfähigkeit bei Sportlern. O&M Ernährung, *Sonderheft Sportmedizin, 11.*
- Wienecke, E. (2018). Regulator für eine optimale mentale und psychophysische Leistungsfähigkeit. *Sportärztezeitung, 2/2018.*
- Wienecke, E. et al. (2020). Meilensteine in der Gesundheitsmedizin: Optimierte Ernährung, individualisierte Mikronährstoffzufuhr und deren Einfluss auf die psychisch/physische und mentale Leistungsfähigkeit. In A. Dreier, R. Merk & B. Seel (Hrsg.), *Meilensteine in der Gesundheitsmedizin.* (Schriftenreihe der FHM). Heft 12.
- Wienecke, E. (2019). Meilensteine in der Regulationsmedizin: Individualisierte Mikronährstoffrezepturen. Case-Reports von 144 Profis. *Sportärztezeitung, 2/2019.*
- Wienecke, E. (2001). *Mineralstoffe, Spurenelemente und Vitamine.* Wissenschaftlicher Abschlussbericht der Screening-Aktion Bertelsmann 2001; Herausgeber Bertelsmann-Stiftung.
- Wienecke, E. & Nolden, C. (2016). *Langzeit-HRV-Analyse zeigt Stressreduktion durch Magnesiumzufuhr,* MMW Fortschritte in der Medizin. Originalien Ili-IV/2016.

- Wienecke, E. (2016). Metabolische Dysfunktionen. Frühzeitige Erkennung im Energiestoffwechsel bei Leistungs- und Spitzensportlern. *Sportärztezeitung, 4/2016.*
- Wormer, E. J. (2014). *Hashimoto.* ISBN 978-3-86374-175-4, Mankau, Murnau,
- Zaalberg, A. P., Nijman, H., Bulten, E., Stroosman, L. & van der Staak, C. (2010). Effects of nutritional supplements on aggression, rule – breaking, and psychopathology among young adult prisoners. *Aggressive Behaivior, 36*, 117-126.
- Zechmann, W. (2007). *Wo beginnt die subklinische Hypothyreose?* https://www.schilddrueseninstitut.at/multimedia/UIM4_Schilddruese.pdf
- Zhu, C., Kobayashi, K. & Loladze, I. et al (2018). Carbon dioxide (CO_2) levels this century will alter the protein, micronutrients, and vitamin content of rice grains with potential health consequences for the poorest rice-dependent countries. *Science Advances, 4, no. 5*, eaaq1012; DOI: 10.1126/sciadv.aaq1012

3 Information online (also for download)

Further information, documents, and materials can be found online:

- https://hepart.ch/fileadmin/user_upload/hepart/PDFs/pdf_Produktinformationen_HCK/Hydro_Cell_key_A4_deutsch.pdf; Letzter Zugriff 05.09.2020
- http://physiologie.cc: Die Website eines österreichischen Physiologen (Prof. em. Dr. Helmut G. Hinghofer-Szalkay, Uni Graz), die komplizierte Zusammenhänge anschaulich erklärt.
- https://foodandmoodcentre.com.au/2016/07/diet-in-pregnancy/; Letzter Zugriff 21.07.2020
- https://salusmed.ch/megatrends-verlangen-nach-personalisierten-gesundheitsloesungen/
- https://www.arte.tv/de/videos/082725-000-A/unser-hirn-ist-was-es-isst/; Letzter Zugriff 21.07.2020
- https://www.dgsm.de/downloads/dgsm/arbeitsgruppen/ratgeber/neu-Nov2011/Schlafstoerung_A4.pdf; Letzter Zugriff 21.07.2020

- https://www.fh-mittelstand.de/fileadmin/pdf/Schriftenreihe/Schriftenreihe_12_web.pdf; Letzter Zugriff 25.07.2020
- https://www.omegametrix.eu/index_patienten.php; Letzter Zugriff 21.07.2020: Informationen zum HS-Omega-3-Index
- https://www.thuenen.de/de/thema/klima-und-luft/experimentelle-klimawirkungsforschung/wie-das-face-experiment-funktioniert/: Johann Heinrich von Thünen-Institut, Bundesforschungsinstitut für Ländliche Räume, Wald und Fischerei, Braunschweig: FACE-Experiment; Letzter Zugriff 21.07.2020
- https://www.youtube.com/watch?v=pDc6m0UrEa4: Interview mit Prof. Dr. Elmar Wienecke auf der Medica in Düsseldorf 2020: Einfluss individualisierter Mikronährstoffe auf Stressoren; Letzter Zugriff 21.07.2020
- http://www.myoreflex.de/media/downloads/Glycoplan_Ampel-Liste_Mosetter.pdf: Glykoplanampel von Dr. Mosetter; Letzter Zugriff 21.07.2020
- https://www.zukunftsinstitut.de: Informationen zu Megatrends
- www.energyforhealth.de: Mikronährstoffanalyse für Jedermann, Personen mit wenig Befindlichkeitsstörungen
- www.fh-mittelstand.de/mikronaehrstofftherapie/
- www.saluto.de: Mikronährstoffanalyse für Führungskräfte, Spitzensportler, Vorerkrankte und Personen mit größeren Befindlichkeitsstörungen
- www.stiftung-mikronaehrstoffe.de

4 Acknowledgments

Successful visionary concepts are only possible with good partners. A special thank you goes out to the responsible professors at the Fachhochschule des Mittelstands (FHM Bielefeld), Prof. Dr. Merk, Prof. Dr. Dreier, and Prof. Dr. Niemeier, who made the establishment of the master's and certification course Micronutrient Therapy and Regulatory Medicine -the only one of its kind in Europe- possible at the technical college.

I would especially like to thank Prof. Dr. Elke Zimmermann. She made it possible for SALUTO to work out of the university in Bielefeld, Germany.

A special thank you also to Dr. Kurt Mosetter MD (Director of the Center for Interdisciplinary Therapies in Constance, initiator of Myoreflex therapy) and his team. He has supported us in recent years with his enormous professional expertise and set the course as lecturer for the master course at the FHM Bielefeld.

I would also like to thank my very committed students (doctors, physical therapists, alternative practitioners, nutritionists, and sports scientists) who, with their innovative ideas, lend optimal support to the development of the master course.

And I would like to thank the Bertelsmann Foundation, particularly Liz Mohn, who made a crucial contribution to the now internationally successful "Energy on Prescription" concept by initiating a screening action with her colleagues.

5 About SALUTO (Society for Sport and Health)

SALUTO was founded in the area of sports medicine at the University Bielefeld/ Germany. Prof. Dr. Elmar Wienecke is the co-founder and owner. A combination of medical services, diagnostics, science, and research has made SALUTO an internationally recognized competence center for health and fitness in Germany. Many research projects and studies are analyzed as part of the master course Micronutrient Therapy and Regulatory Medicine in cooperation with the Fachhochschule des Mittelstands (FHM) technical college in Bielefeld/Germany.

To date more than 60,000 tested individuals have been included as *case reports* in the world's largest micronutrient database. It includes 10,542 target values/reference values of micronutrient concentrations (results from empirical long-term research projects/studies) in 297 categories with the respective cluster (gender, age, preexisting conditions, athletic activity).

The "Energy on Prescription" concept sprang from clinical studies and many research projects. The goal of this successful energy concept: The timely detection and correction of biochemical disruptions to guarantee good health and performance capacity. Meanwhile Olympic champions, world champions European champions, and German champions are benefitting from this holistic concept. But many executives have also been successfully using this concept for years.

6 About the Foundation for Micronutrients - Prevention, Health, Quality of Life

Foundation for Micronutrients, Prevention, Health, Quality of Life

"Many overload responses and the resulting disorders can verifiably be avoided with an optimal energy intake."

However, to date the scientific community discounts these relationships.

According to statements from international scientists, there is "little knowledge" about the positive effects of a targeted micronutrient intake. It is all still in its infancy. After 25 years of meticulous scientific work and the now more than 60,000 compiled case reports, Prof. Dr. Elmar Wienecke, sports scientist and founder of SALUTO the Competence Center for Health and Fitness, felt compelled to establish the Foundation for Micronutrients, Prevention, Health, Quality of Life.

New supplementary and alternative means in the area of micronutrient therapy are being researched and find practice-oriented application by a team of physicians, sports scientist, biologists, natural scientists and nutritionists.

Foundation goals

Prevention

- Establishing micronutrient therapy in a demographically aging population with respect to age, gender, lifestyle, preexisting conditions.

Research

- Early detection of biochemical disruptions.
- Practical application in the area of molecular biological micronutrient therapy in active people (recreational, competitive and elite sports).

- Creating discussion forums, significance of the micronutrient intake in complementary medicine.
- Promoting education with respect to micronutrient therapy by informing the population and transferring basic knowledge about how micronutrients can really help. Integrating basic knowledge into the training structure of physicians, physical therapists, and nutritionists.

Publications and studies

Establishing the master's and certification course Micronutrient Therapy and Regulatory Medicine at the FHM Bielefeld technical college – research assignments in the area of exercise- and micronutrient therapy. Long-term goal: Integrating contents into physicians' medical licensure act.

Research in this area is intended to ensure early detection of biochemical disruptions so they can be counteracted with an optimal energy and micronutrient intake.

For you

If you are interested in supporting the concept and the work of the Foundation for Micronutrients, Prevention, Health, Quality of Life, we would love for you to contact us.

The foundation pursues exclusively non-profit objectives.

Foundation Chairperson: Prof. Dr. Elmar Wienecke

www.stiftung-mikronaehrstoffe.de

Email: stiftungmikronaehrstoffe@t-online.de

7 About Energy for Health

While SALUTO offers customized micronutrient problem solutions for pre-diseased persons, competitive athletes, or people with a need for specific advice, Energy for Health (https://energyforhealth.de) addresses people looking for an easy way to obtain an individualized micronutrient analysis with a customized nutrient recommendation. The demand-oriented concept -optimizing performance or activating the metabolism, optimizing wellbeing or improving sleep behavior- supported by a customized and optimal micronutrient supply.

The test box can be easily ordered online and blood and urine samples as well as the enclosed anamnesis form are sent back. The lab results are then entered into the globally unique database with comparable values from more than 60,000 persons. The results contained therein are compared to the personal data via different parameters. The competence team creates and verifies a suggested formulation. Only then does the customer receive the recommendation for the customized micronutrient formulation.

8 Credits

Cover image:	© AdobeStock, © Saluto
Cover design:	Sannah Inderelst
Interior design:	Sannah Inderelst
Layout:	Anja Elsen
Interior photos:	Chapter openers: © AdobeStock, © AdobeStock: pp. 27, 35, 39, 48, 68, 111, 127; © Thinkstock: pp. 50, 51, 52, 54, 94, 101; © Bittium: p. 61; © Henz et al.: pp. 74, 75, 76; © Saluto: pp. 38, 227; © Hepart, Schweiz: p. 141; © Elmar Wienecke: pp. 87, 88
Figures:	Reproduced by Satzstudio Hilger
Copyeditor:	Joshua Brazee
Translator:	AAA Translation, www.AAATranslation.com